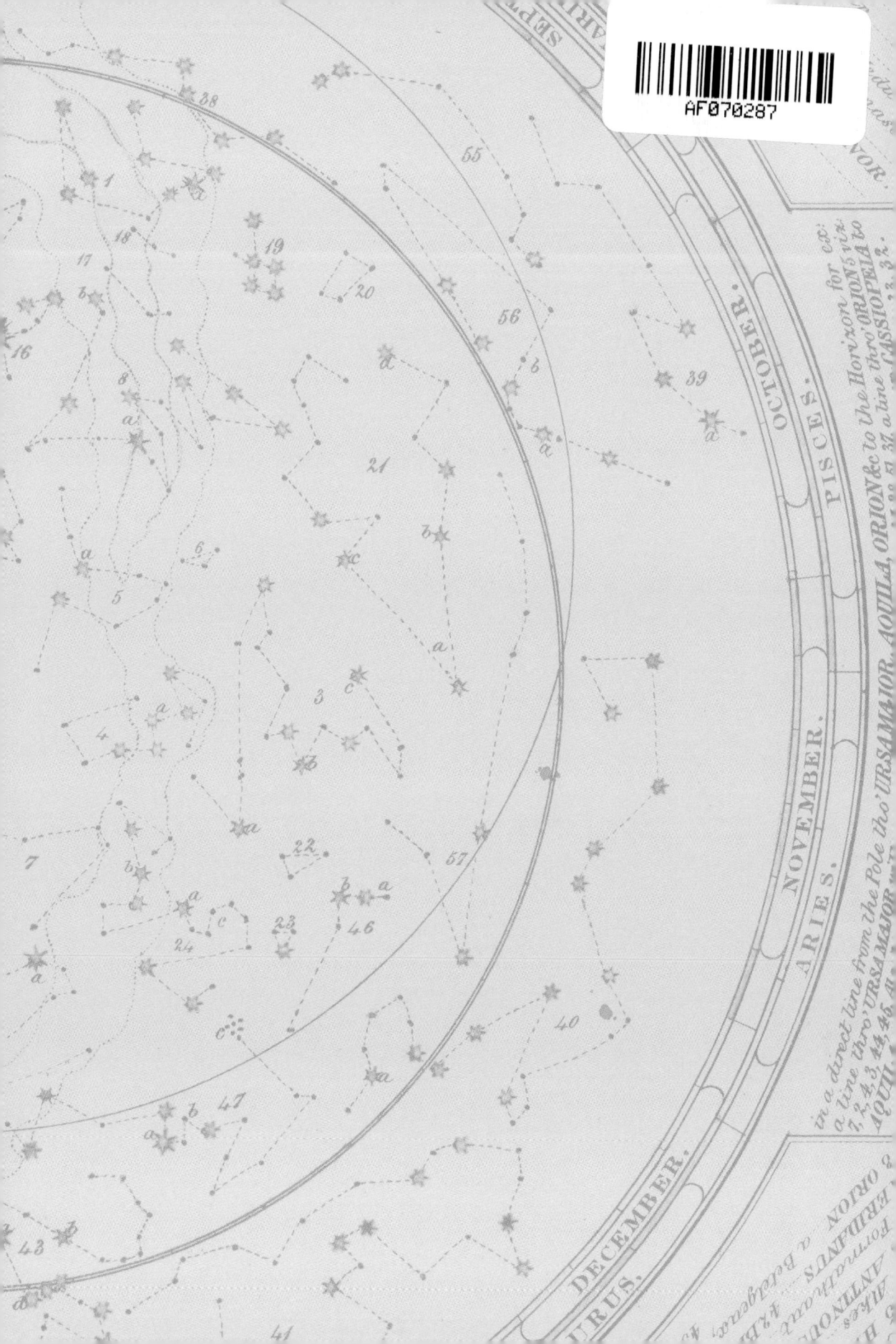

**ROYAL OBSERVATORY GREENWICH**

A HISTORY IN OBJECTS

# Royal Observatory Greenwich

## A History in Objects

LOUISE DEVOY

# Contents

| | |
|---|---|
| Foreword | 9 |
| Introduction | 10 |
| The Astronomers Royal | 12 |
| The quest for longitude | 14 |

## JOHN FLAMSTEED (FIRST ASTRONOMER ROYAL, 1675–1719) — 20

1. Portrait of Charles II — 22
2. Print of *Prospectus intra Cameram Stellatam* — 24
3. Year-going clock movement — 26
4. Flamsteed's 7ft equatorial sextant — 28
5. *The Royal Observatory from Crooms Hill* — 30
6. 1714 Longitude Act — 32
7. *Historia Coelestis Britannica* (1725) — 34
8. *Atlas Coelestis* (1729) — 36

## EDMOND HALLEY (SECOND ASTRONOMER ROYAL, 1720–42) — 40

9. Chart of the southern celestial hemisphere — 42
10. Halley's transit telescope — 44
11. Halley's 8ft mural quadrant — 46
— **In focus:** How does a mechanical clock work? — 48
12. Regulator no.1 by Graham — 50
13. Marine timekeeper by Harrison, 'H1' — 52
14. Marine timekeeper by Harrison, 'H2' — 54

## JAMES BRADLEY (THIRD ASTRONOMER ROYAL, 1742–62) — 56

15. Bradley's 12.5ft zenith sector — 58
16. Bradley's 8ft mural quadrant — 60
17. Bradley's transit telescope — 62
18. Marine timekeeper by Harrison, 'H3' — 64

## NATHANIEL BLISS (FOURTH ASTRONOMER ROYAL, 1762–64) — 66

## NEVIL MASKELYNE (FIFTH ASTRONOMER ROYAL, 1765–1811) — 68

19. *Nautical Almanac* (1767) — 70
20. Maskelyne's observing suit — 72
21. Marine timekeeper by Harrison, 'H4' — 74
22. Pair of floor globes — 76
23. Longitude calculation sheet — 78
— **In focus:** Transit of Venus expeditions — 80

| | | |
|---|---|---|
| 24. | Gregorian telescope | 84 |
| 25. | Maskelyne's Copley Medal | 86 |
| 26. | Chronometer no.36 by Arnold | 87 |
| 27. | Sextant by Ramsden | 88 |
| —— | **In focus:** London's community of clock and scientific instrument makers, 1750–1800 | 90 |
| 28. | Sophia Maskelyne's dress | 92 |
| 29. | Portrait of Margaret Maskelyne | 94 |
| 30. | Caroline Herschel's dress and bonnet | 96 |
| 31. | *Astronomical Observations* (1798) | 98 |
| 32. | Model of a chronometer escapement | 100 |
| 33. | Print of the Easter Festival | 102 |
| 34. | Print of the view from the camera obscura | 103 |

## JOHN POND (SIXTH ASTRONOMER ROYAL, 1811–35) — 106

| | | |
|---|---|---|
| 35. | Pond's mural circle | 108 |
| 36. | Astronomer's alarm clock | 110 |
| 37. | Rain gauge | 111 |
| 38. | Chronometer no.427 by Brockbanks | 112 |
| 39. | Time ball | 114 |

## GEORGE BIDDELL AIRY (SEVENTH ASTRONOMER ROYAL, 1835–81) — 116

| | | |
|---|---|---|
| 40. | Engraving of Cambridge Observatory | 118 |
| 41. | Sketch of Playford cottage | 120 |
| 42. | Airy's hole punch | 122 |
| 43. | Grand orrery | 123 |
| 44. | Regulator no.587 by Dent | 124 |
| 45. | *Daylight View of the Great Comet of 1843* | 126 |
| —— | **In focus:** The Age of Magnetism | 128 |
| 46. | The Magnet House | 132 |
| 47. | Airy Transit Circle | 133 |
| 48. | The fireball meteor of 1850 | 136 |
| 49. | Shepherd Motor Clock and Gate Clock | 138 |
| 50. | British local time map | 140 |
| 51. | Marine chronometer no.10 by Barraud | 142 |
| 52. | Public imperial standards of length | 144 |
| 53. | Magnetogram of the Carrington Event on 1 September 1859 | 145 |
| 54. | Airy's dip circle | 146 |
| 55. | Hourly time signal relay | 148 |
| 56. | Single-needle telegraph | 149 |
| —— | **In focus:** Meteorology becomes a science | 150 |
| 57. | Public barometer | 154 |
| 58. | Earth current galvanometer | 156 |
| 59. | The Airy family | 158 |
| 60. | *The Midnight Sky* (1869) | 159 |
| —— | **In focus:** Cable production in Greenwich | 160 |
| 61. | Presentation box of submarine cable samples | 164 |
| 62. | Airy's Sèvres vase | 166 |
| 63. | Imperial Order of the Rose | 168 |
| 64. | Dallmeyer photoheliograph | 170 |
| 65. | Test plate for the Janssen photographic revolver | 172 |

| | | |
|---|---|---|
| 66. | Sunshine recorder | 173 |
| | **In focus:** Splitting starlight with spectroscopy | 174 |
| 67. | Two-prism spectroscope | 178 |
| 68. | Kew-pattern unifilar magnetometer | 182 |

## WILLIAM HENRY MALONEY CHRISTIE (EIGHTH ASTRONOMER ROYAL, 1881–1910) — 184

| | | |
|---|---|---|
| 69. | The 1884 International Meridian Conference | 186 |
| 70. | World time converter | 188 |
| 71. | Astrographic telescope | 192 |
| | **In focus:** The women of the Carte du Ciel | 194 |
| 72. | Solar plate micrometer | 198 |
| 73. | 'Greetings from Greenwich' postcard | 200 |
| 74. | Pocket chronometer 'Arnold 485' | 201 |
| 75. | 28in. Great Equatorial Telescope | 202 |
| 76. | *The Secret Agent* (1907) | 206 |
| 77. | Altazimuth Pavilion | 208 |
| 78. | New Physical Observatory | 210 |
| | **In focus:** Sports and social clubs | 212 |
| 79. | Spider fork | 214 |
| 80. | Article by Maunder about the 'canals' on Mars | 215 |
| 81. | *The Heavens and their Story* (1908) | 216 |
| 82. | 30in. Thompson Reflector | 218 |

## FRANK WATSON DYSON (NINTH ASTRONOMER ROYAL, 1910–1933) — 220

| | | |
|---|---|---|
| 83. | Spherical plate calculator for star coordinates | 222 |
| 84. | Henry Outhwaite, Observatory Secretary | 224 |
| 85. | First World War binoculars | 226 |
| 86. | Double star catalogue by Jonckheere | 228 |
| 87. | Glass photopositive of the 1919 total solar eclipse | 230 |
| 88. | Measuring device for the star trail camera plates | 232 |
| 89. | Regulator no.2016 by Dent | 234 |
| 90. | Shortt free-pendulum clock system | 236 |
| 91. | Glass photopositive of the Magnetic Pavilion at Abinger | 238 |
| 92. | Observer's card for the Giggleswick total solar eclipse | 240 |
| 93. | Warren Synclock | 242 |
| 94. | Rupert Gould's notebooks | 244 |
| 95. | Mary French with the plate micrometer | 246 |

## HAROLD SPENCER JONES (TENTH ASTRONOMER ROYAL, 1933–1955) — 248

| | | |
|---|---|---|
| 96. | Slitless spectrograph | 250 |
| 97. | Occultation machine | 252 |
| 98. | Philip Laurie's wartime diaries | 254 |
| 99. | Quartz clock and frequency standard | 256 |
| 100. | HP5061A caesium-beam atomic clock | 258 |

| | |
|---|---|
| Epilogue | 260 |
| Glossary | 268 |
| Improvements in accuracy over time | 270 |
| Additional readings, references and quotations | 273 |
| Acknowledgements | 287 |
| Image credits | 288 |

# Foreword

The history of the Royal Observatory, Greenwich is a story of exploration and curiosity. In 1675 when Flamsteed laid the Observatory's foundation stone, it was so he could catalogue the stars to define time itself. In doing so, he paved the way for mariners to confidently sail across the entire globe. Halley, Bradley, Bliss and Maskelyne followed, opening our horizons to new wonders in our diverse Solar System, while continually refining the measurement of time to 'perfect the art of navigation'. Fast forward to 1919 and suddenly our whole understanding of time and space has been turned on its head by Dyson, the ninth Astronomer Royal. By observing stars during a total solar eclipse, his Greenwich team proved Einstein right. Time and space are not absolute: they are warped by massive objects like our own Sun. Soon after we learnt that our Milky Way isn't alone in the Universe, that the faint fuzzy objects seen through the Observatory's telescopes were in fact distant galaxies. Over the course of 350 years, our ambitions to explore the Earth have evolved into a desire to explore the entire Universe, and historian and astrophysicist Dr Louise Devoy shows us how the Astronomers Royal at Greenwich were key architects in this mission. This book tells the stories of the eccentric characters and technical innovation that placed Greenwich at the centre of the world, beautifully illustrated by unique artefacts and historical records from the Museum's collection.

Today the title of Astronomer Royal, and the sister role of Astronomer Royal for Scotland that I hold, is purely honorary. We no longer have the pleasure of living in opulence within the stunning Observatory grounds or hosting sumptuous high society dinner parties. But our thirst for knowledge and understanding of the world around us is undiminished. Thanks to the foresight of our more recent Astronomer Royal predecessors, scientific observations have moved from the light polluted and often clouded cities of London and Edinburgh to telescopes on remote arid dry mountaintops, and even in space. By collaborating with international partners to develop observatories all around the world, the work of the two Royal Observatories led to the emergence of global scientific institutions. Flamsteed would surely be astonished by all the varied and groundbreaking scientific discoveries that were seeded by his 1675 vision of a Royal Observatory.

**Professor Catherine Heymans**
Astronomer Royal for Scotland
University of Edinburgh

# Introduction

*'I cannot but feel a satisfaction in thinking that the Royal Observatory is thus quietly contributing to the punctuality of business through a large portion of this busy country.'*

George Biddell Airy,
Seventh Astronomer Royal, 1853

Observatories cast our minds upwards to the wonders of the cosmos but the objects featured in this book remind us that astronomy is a challenging, human endeavour that relies on a multitude of people, ideas and technologies. Each object lies at the centre of a complex web of hidden history and provides us with a window into the work of the Astronomers Royal at Greenwich, who dutifully persevered to map the stars and measure time to help us navigate around the globe.

Since its foundation 350 years ago, the Royal Observatory in Greenwich has always traded in data: vast amounts of stellar, lunar and planetary positions and timings, all of which depended on specially designed buildings, clocks and telescopes, managed by a growing army of assistants, technicians and human computers. These tasks relied on tools and equipment that were carefully maintained, adjusted and repaired; some were even repurposed and modified to extend their working lives for decades.

Technological progress is a tangible thread that is woven throughout these stories as we see clocks becoming more accurate and telescopes morphing into giant machines. We can also see how the adoption of new instruments and techniques from laboratories to observatories in the mid-1800s transformed astronomy into astrophysics: astronomers could now analyse the stars, rather than simply measuring their positions for timekeeping. At the same time, the Observatory's staff became involved in new areas of scientific investigation such as studying the Earth's magnetic field, taking daily photographs of the Sun and collecting weather readings. Alongside the technological marvels of highly accurate timekeepers and sophisticated telescopes, we can also spot the objects that triggered debate, dispute and controversy. Medals and presentation gifts convey stories of outstanding technical and intellectual achievement while others bear witness to stories of frustration, failure and abandonment.

More broadly, many of these objects highlight the Observatory's role as a networking hub. From informal dinner parties held on-site to formal discussions at universities and learned societies, each successive Astronomer Royal collaborated with the best intellectuals and craftsmen of their time. Similarly, the publication and circulation of the Observatory's data among major institutions across Europe and North America kept Greenwich at the forefront of research. Some of the featured objects are survivors from large-scale collaborative projects that transformed our understanding of the Universe, from expeditions to witness eclipses and transits of Venus through to the Carte du Ciel, an ambitious global project to map the stars using photography.

Other objects demonstrate how the Astronomers Royal were acutely aware of their responsibility to share the Observatory's work with the public, most famously via its rooftop

*Flamstead [Flamsteed] House*, Thomas Hosmer Shepherd and Rudolph Ackermann, 1824 (PAD2226)

time ball, Shepherd Gate Clock and 'six pips' radio time signal. As public interest in astronomy grew during the late 1800s, several Greenwich astronomers shared their passion for the subject through public lectures, articles and books, a trend that continues today.

Behind closed doors, the Observatory was a very private family home for the Astronomers Royal, their families, servants and pets. A series of domestic items helps us explore our curiosity about what it was like to live in such a unique family home. We find personal stories of childhood, romance, faith, illness and grief running in parallel with professional and social responsibilities such as welcoming dignitaries, taking in refugee astronomers and hosting garden parties for charity.

At the opposite end of the scale we find objects that convey the global significance of the Observatory's work. From the 1820s, chronometers purchased by the Admiralty were tested against the Observatory's accurate clocks set by the stars, before being issued to vessels that explored, mapped and defined the British Empire and beyond. Similarly, telegraph cables alongside the railways and under the seas enabled Greenwich astronomers to use these signals to check their longitude with other observatories, all of which contributed to the international decision to designate the Greenwich meridian as Prime Meridian of the World, 0° longitude, in 1884.

Finally, we can trace the Observatory's twilight years in the first few decades of the twentieth century by examining the objects that illustrate how timekeeping technologies shifted from observatories to laboratories with the emergence of quartz and then atomic clocks. The suitability of the Greenwich site itself was also under scrutiny: once surrounded by miles of countryside, the Royal Observatory was now enveloped by London's factories, railway lines and power stations that filled with air with sooty, coal-fuelled smoke. With a declining role in timekeeping and little opportunity to observe the stars, from 1948 the astronomers started to relocate the instruments to a new site at Herstmonceux in East Sussex. They continued their work at the renamed 'Royal Greenwich Observatory' (RGO) until 1990 when operations were relocated once again to Cambridge. By 1998, when the RGO finally closed its doors, astronomers now preferred to use bigger telescopes in better climates abroad. Today, the title of Astronomer Royal is honorary, awarded to an eminent scientist whose role it is to advocate for astronomy both within government and the public sphere.

The historic site of the Royal Observatory at Greenwich became part of the National Maritime Museum in the 1950s and we continue to preserve the stories of the buildings, instruments and people who made this the most famous observatory in the world.

# The Astronomers Royal

The role of Astronomer Royal was originally created to map the stars and measure time by the Sun at the Royal Observatory, Greenwich. Today the position is awarded to a prominent British scientist who is required to advise the monarch on astronomical matters and to promote public engagement with astronomy and space exploration.

**JOHN FLAMSTEED**
1675–1719

Compiled a catalogue of stars that was larger and more accurate than any so far produced.

**EDMOND HALLEY**
1720–42

Famous for his work on comets and his support of Newton's book on gravitation, the *Principia Mathematica*.

**JAMES BRADLEY**
1742–62

Discovered the Earth's wobble on its axis, proved that our planet moves through space and established a new Greenwich meridian.

**NATHANIEL BLISS**
1762–64

Succeeded his friend James Bradley as Astronomer Royal but only lived for another two years.

**NEVIL MASKELYNE**
1765–1811

Focused his efforts on producing the *Nautical Almanac*, a book of astronomical data to help navigators determine their longitude at sea.

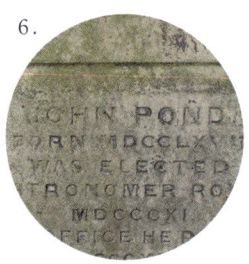

**JOHN POND**
1811–35

An accomplished observer who replaced the Observatory's old equipment with better instruments. Also set up the time ball.

**GEORGE BIDDELL AIRY**
1835–81

A keen innovator who designed new instruments, reorganised the Observatory's work and improved the distribution of accurate Greenwich time for public services.

**WILLIAM CHRISTIE**
1881–1910

An ambitious man who expanded the Observatory's work by introducing more buildings, instruments, staff and collaborative projects.

**FRANK DYSON**
1910–33

Responsible for expanding the distribution of Greenwich time via radio signals and for the organisation of solar eclipse expeditions overseas.

**HAROLD SPENCER JONES**
1933–55

The last Astronomer Royal to live in Flamsteed House. He oversaw the relocation of the Observatory to its country site at Herstmonceux, East Sussex.

**RICHARD VAN DER RIET WOOLLEY**
1956–71

Optical astronomer and influential leader who oversaw observatories and telescope-building programmes in Britain, Australia and South Africa.

**MARTIN RYLE**
1972–82

Wartime radar scientist and inventive physicist who linked multiple radio telescopes to detect faint sources from the early Universe.

**FRANCIS GRAHAM-SMITH**
1982–90

Radio astronomer and student of Ryle who detected the first example of a radio galaxy, Cygnus-A, and helped establish a multinational observatory in La Palma, Canary Islands.

**ARNOLD WOLFENDALE**
1991–95

Particle physicist who invented new detectors for measuring cosmic radiation using satellites, particle accelerators and experimental apparatus within deep mines.

**MARTIN REES**
1995–2025

Cosmologist specialising in black holes and galaxy formation who now investigates long-term global issues such as climate change and asteroid impacts.

**MICHELE DOUGHERTY**
2025–present

Space physicist who is leading on unmanned exploratory missions to Saturn and Jupiter and specialises in studying planetary magnetic fields.

# The quest for longitude

On 10 August 1675, the first Astronomer Royal, John Flamsteed, laid the foundation stone of the new observatory that was starting to emerge on the site of Duke Humphrey's Tower in Greenwich Park. He had a daunting task ahead of him: many nights of observations to create better star charts for 'perfecting the art of navigation', as commissioned by King Charles II (1). By the 1600s there was a growing impetus for mariners to travel further and faster as new, highly lucrative trade routes emerged in Asia, Africa and the Americas. Traders and governments alike were keen to capitalise on the burgeoning European demand for gold, tea, coffee, spices, ceramics and timber, products that often relied on the parallel trade in enslaved people. As seafarers ventured beyond familiar coastlines, they depended on simple instruments and a basic knowledge of the stars to help them navigate across the featureless ocean on lengthy and perilous voyages. The establishment of the Royal Observatory, Greenwich – Britain's first state-funded scientific institution – was a clear indication of the government's interest in taking advantage of the economic and political opportunities of better global navigation.

### Latitude but not longitude
Any location on Earth can be defined as a set of coordinates based on horizontal lines of latitude (north–south position) and vertical lines of longitude (east–west position, see *Glossary*). Finding latitude in the 1600s was relatively easy as mariners could use either the height (altitude) of the Sun above the horizon at noon, or else measure the height of the north star, Polaris, at night. Voyagers in the southern hemisphere could also rely on the constellation of the Southern Cross (Crux). But longitude was much more difficult, especially as there was no natural equivalent of the equator – 0° latitude – to indicate where longitude begins and ends. Instead, astronomers realised that they would have to utilise the Earth's daily rotation on its axis as a measure of longitude, whereby our planet rotates 360° every 24 hours, or 15° per hour. If two observers see the same event in the sky, such as an eclipse, but their local times are three hours different then they must be separated by 45° degrees longitude. But with no direct communication possible in the 1600s, how could they compare their observations simultaneously? It was a challenging question that would occupy the minds of the best astronomers, mathematicians and scientific instrument makers for centuries to come.

### Measuring longitude through time difference using astronomy
Flamsteed's work was naturally focused on creating an astronomical solution for measuring longitude at sea. One such approach was to use the moons of Jupiter as a celestial clock.

⬆
*Paris Observatory*, Gabriel Perella, after 1729 (PAJ3502)

⬇
Replica of a Galilean telescope, unknown maker, 1974 (AST0926)

In 1610, the Italian astronomer Galileo Galilei turned his telescope towards to the giant planet and noticed four dots of light that appeared to change position each night, seemingly disappearing and then reappearing from behind Jupiter's disc. Within two years, he had made enough observations to predict the configuration of these moons as seen from his hometown of Florence, meaning that another observer watching Jupiter elsewhere could potentially determine their time difference and longitude from the Tuscan city. Galileo submitted his idea to the Spanish and Dutch governments, who were offering generous cash rewards for longitude solutions, but neither country explored his proposal in detail.

In Paris, the 'Sun King', Louis XIV, was persuaded by his ministers to invest in astronomy and the resulting observatory opened in 1667 under the directorship of the Italian-born astronomer Giovanni Domenico Cassini. One of Cassini's first tasks was to re-survey the outline of France using longitude measurements based on observations of Jupiter's moons. The method worked well, leading the King to claim that he had lost more land to his astronomers than to his enemies! Decades later, Flamsteed dutifully recorded his own observations of Jupiter's moons from Greenwich and shared his data with Cassini but the technique was rarely used at sea as navigators struggled to make precise observations with a telescope from a moving ship.

As decreed by Charles II in 1675, Flamsteed's primary task was to create an improved catalogue of star positions to help develop the so-called 'lunar distance method'. Nearly

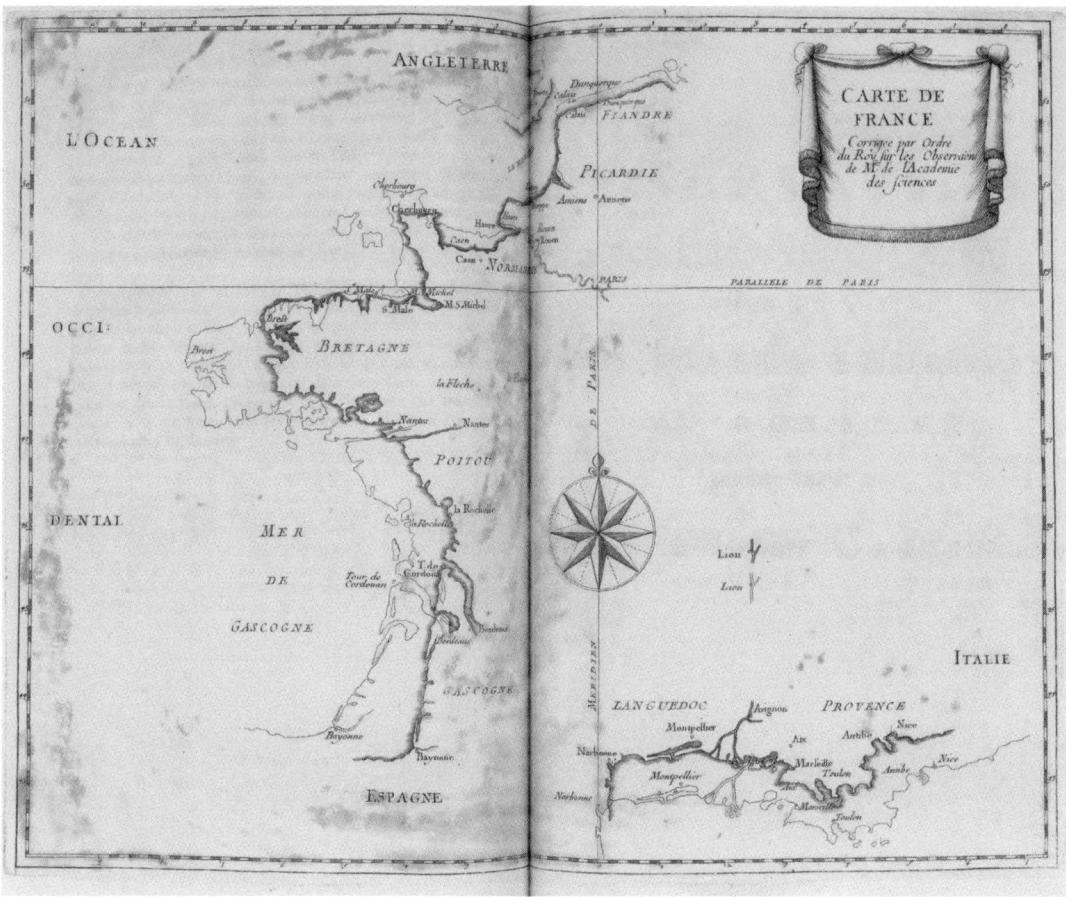

200 years before the Observatory's founding, a German scholar called Johann Werner had proposed using the angle (distance) between the Moon and certain bright stars as a means of comparing observations to determine longitude. Each night, the Moon appears to move relative to the background stars. In theory, if two observers at different locations know when the same angle is visible between the Moon and a certain star, they could calculate their longitude from the time difference. In reality, it would take another three centuries for scholars to create better star charts, design more accurate instruments and develop more sophisticated mathematics to turn Werner's idea into a practical technique.

**Measuring longitude through time difference with a sea clock**
Astronomers were not the only scholars interested in solving the longitude problem. In 1530, the Dutch mathematician Gemma Frisius proposed taking a clock to sea, set to home time, to use as a comparison against local time. He assumed that 'while we are on our journey we should see to it that our clock never stops'. But the clocks of his time were indeed liable to stop and could lose or gain several minutes per day – equivalent to several miles of longitude – that made them unsuitable for oceanic voyages. A century later, Galileo wrote about the good timekeeping properties of a pendulum that swings in equal intervals of time (isochronous), but it was the Dutch scholar Christiaan Huygens who eventually transformed the concept into the first working pendulum clock in 1657. Although they were the most accurate clocks of the day and became an essential part of observatory timekeeping, these instruments were wholly unsuitable for use on rocking ships that disrupted the pendulum's swing. While various clockmakers proposed different ideas, it would take another century before the Lincolnshire carpenter John Harrison

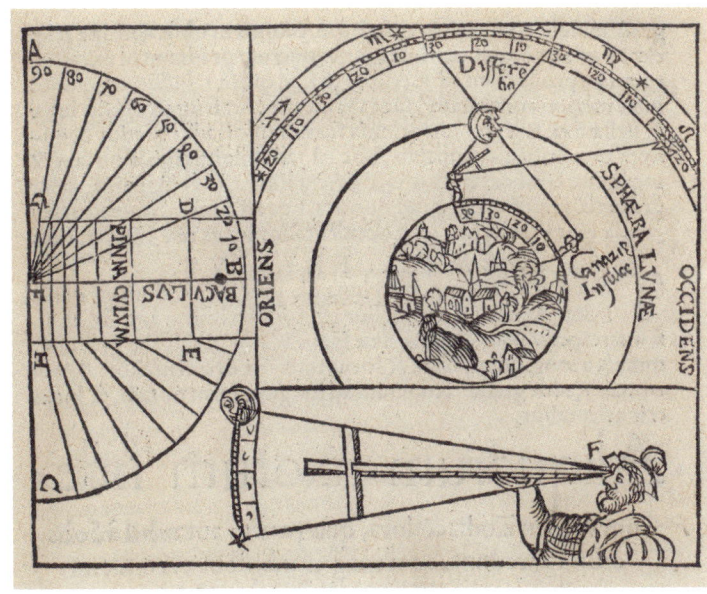

A new map of France from *Recueil d'Observations*, 1693. The difference between the faint and bold outline shows how much territory was 'lost' when astronomers used Jupiter's moons for measuring the longitudes of coastal towns (PBG0584)

Two observers on different sides of the Earth use an instrument called a cross-staff to measure the angle ('Difference' shown opposite) between the Moon and a nearby star on the ecliptic (the apparent path of the Sun, Moon and planets), illustrated in Peter Apian's *Cosmographia*, 1524 (PBD1225)

An early experimental marine timekeeper, unsigned, early 1660s (ZBA6944)

(overleaf)
*The East Prospect of Dr Flamsteads [Flamsteed's] House in Greenwich Park*, Jean Simon and Henry Overton, about 1720 (PAJ2139)

created the first practical 'marine timekeeper', or sea clock – 'H4' (21) – that ushered in a new era of chronometer-based navigation.

**Working together**
Although the Royal Observatory was ostensibly established to solve the longitude problem using astronomy, it played a vital role for over two centuries in bringing together both astronomical and horological approaches to measuring longitude as time difference. The main components of the lunar distance method were all dependent on contributions by the Astronomers Royal and their assistants, from improved star catalogues (7), compiling the *Nautical Almanac* (19), testing lunar data from European colleagues (16) and advising on the design of the sextant (27). Similarly, the Observatory was responsible for testing and rating thousands of Admiralty chronometers (51), providing GMT for chronometer makers (39, 74)

and issuing time signals via telegraph networks (49). Ultimately, these factors all contributed to the Observatory's designation as Prime Meridian of the World – 0° longitude – in 1884 (69).

JOHANNES FLAMSTEEDIUS Derbiensis
Astronomiæ Professor Regius. Anno Ætatis 74 Obijt
Decem: 31 1719

FIRST ASTRONOMER ROYAL, 1675–1719

# John Flamsteed
## 1646–1719

Son of a Derbyshire businessman, John Flamsteed was a sickly child who left grammar school at the age of 15. Between bouts of illness, he taught himself mathematics, geometry and practical astronomy using books kindly lent by local scholars. In 1670 he travelled to London where he met Jonas Moore, Surveyor-General of the Ordnance, who was responsible for mapping and building fortifications on behalf of the Crown. Moore cultivated the young astronomer through introductions to helpful contacts and paying for new astronomical instruments. When King Charles II authorised the Royal Society's recommendation for the construction of a new observatory in 1675, Moore's protégé was an obvious choice for the new 'astronomical observator'. The Surveyor-General continued to provide much financial and logistical support to Flamsteed until his own death in 1679.

Once installed at the new observatory, Flamsteed began a systematic programme of mapping the stars with state-of-the-art instruments and timekeepers, becoming the first major astronomer to use telescopic sights to make his observations more accurate. He also took daily measurements of the Sun to create Greenwich Mean Time, measured the Earth's rotation and used his knowledge of the Moon's orbit to improve tidal predictions for ferrymen on the Thames.

In October 1692 he married Margaret Cooke, daughter of a London lawyer, who became self-taught in mathematics and astronomy. A tantalising comment in Flamsteed's notes – 'solus cum sponsa' [alone with wife] – suggests that she was indeed present during some observations but her exact involvement with his work is unclear. She seemingly made a good impression as a hostess, with the diarist Samuel Pepys writing to thank Flamsteed in July 1697 for '[a] Crowd of Favours, both Intellectual and Culinary, at my late Visit ... heighten'd, by the Conversacion and Kindness of your Excellent Lady'.

As the years passed, Flamsteed's peers became exasperated by his delay in preparing his observational work for publication, leading to a now infamous dispute with Isaac Newton and Edmond Halley in 1712. After Flamsteed's death in 1719, Margaret and the Observatory's assistants worked hard to ensure that his catalogue of 3,000 stars, plotted with unparalleled accuracy, would cement Flamsteed's legacy as an accomplished practical astronomer.

*John Flamsteed*, George Vertue and Thomas Gibson, 1721 (PAD2713)

1

## PORTRAIT OF CHARLES II

Prominently positioned above the door of the Observatory's main room (2), this portrait by the King's Principal Painter served as a powerful symbol of the institution's royal prestige and patronage. In 1674, members of the Royal Society began to explore the possibility of constructing an observatory. But their plans were disrupted by the arrival of a French astronomer, Sieur de Saint-Pierre, who was introduced to the English Court by the Duchess of Portsmouth, fellow Breton Louise de Kéroualle. Saint-Pierre claimed that he had improved a version of the lunar distance method of finding longitude at sea (see *The quest for longitude*), prompting Charles II to appoint members of the Royal Society to assess the claims. The Fellows handed over the data to Flamsteed, who was already well known for his observational skills and mathematical prowess. By the end of February 1675, Flamsteed had concluded that Saint-Pierre's technique was fundamentally flawed.

Armed with this result, a delegation from the Royal Society persuaded the King that England needed an observatory to produce better star charts to improve navigation at sea, hopefully saving lives and improving traders' profits. On 4 March 1675, the King signed a Royal Warrant appointing Flamsteed as 'astronomical observer' with the objective of 'perfecting the art of navigation'. The Derbyshire astronomer started his observations straight away from his lodgings at the Tower of London, thanks to his patron, Jonas Moore, Surveyor-General of the Ordnance, whose organisation would fund the new scientific institution for the next 150 years. Meanwhile, the Royal Society assessed potential observatory locations in Hyde Park and Chelsea before deciding upon the ruins of Greenwich Castle, freely available from the Crown. With its hilltop location, away from the smoke of the city but still accessible by river and road, the site was the preferred choice of the architect Christopher Wren and was confirmed by Royal Warrant on 22 June 1675. The foundation stone was laid six weeks later and Flamsteed moved in the following year. Despite his initial input, the King seemingly never came to visit his observatory, relying instead on Flamsteed's visits to Whitehall for updates.

*Charles II*, Sir Peter Lely, about 1670 (BHC2609)

↑
*Prospectus intra Cameram Stellatam* [View inside the Star Room], Francis Place, 1712 (after Robert Thacker, 1676) (ZBA1808)

# PRINT OF *PROSPECTUS INTRA CAMERAM STELLATAM*

Famous for his work restoring London after the disastrous Great Fire of 1666, the architect Christopher Wren was commissioned to design the new observatory, assisted by the Surveyor to the City of London, Robert Hooke. Based on the fashionable geometric trends of the Baroque period, the Observatory building consisted of a square-shaped ground floor with a series of 'dwelling rooms', topped by an impressive octagonal 'Star Room' upstairs. The foundation stone was laid on 10 August 1675 and Flamsteed made his first observation from this room on 31 May 1676.

As one of the instrumental figures behind the creation of the Observatory, Flamsteed's sponsor Jonas Moore wanted to immortalise his achievement and so he commissioned engraver Francis Place to create a set of twelve etchings depicting the new institution and its instruments, based on drawings made by Robert Thacker. This view of the Star Room shows an astronomer using a quadrant (quarter-circle) to measure the height (altitude) of the stars above the horizon. The open doors leading out onto the balcony provide us with a glimpse of ships meandering along the River Thames. On the right, another observer uses a ladder to support a long telescope tube. Astronomers struggled with these unwieldy instruments until the development of achromatic lenses (5) around 1730, which enabled them to create shorter telescopes with less colour distortion.

In the centre of the image, the portraits of Charles II (1, left) and James II (right) remind us of the Observatory's royal patronage. The wall cavity behind Charles's portrait originally housed two pendulums, each one 13ft (4m) long, for the highly accurate clocks (3) whose dials can be seen to the left of the door.

Despite its aesthetic charms, the Star Room – known today as the Octagon Room – was superseded by new buildings and instruments installed elsewhere on site. Fifth Astronomer Royal Nevil Maskelyne made the last observations here during the 1790s and over successive centuries the space has been used as a meeting room, calculating room, coroner's court, museum and concert hall. It was opened to the public by Prince Philip, Duke of Edinburgh, on 8 May 1953.

## 3
## YEAR-GOING CLOCK MOVEMENT

Returned to the national collection after 250 years of private ownership, this clock movement is one of a pair that originally ticked away behind the panelled walls of the Octagon Room (2). As the Observatory building began to take shape in late 1675, Flamsteed's patron, Jonas Moore, ordered these highly accurate timekeepers from Thomas Tompion, the best clockmaker in London at the time.

Flamsteed needed such state-of-the-art timekeepers to prove that the Earth rotates on its axis at a constant rate (isochronal), a key assumption in all astronomical methods for measuring longitude (see *The quest for longitude*). First, he improved his existing 'Equation of Natural Days' to smooth out variations in the length of the apparent solar day, caused by the tilt of the Earth's axis and the elliptical shape of its orbit. By formulating an average or 'mean' day of exactly 24 hours, he set these clocks to show this new 'Greenwich Mean Time' (GMT). Next, he set up a 6ft (1.8m) telescope on the western wall of the Octagon Room to measure the time interval between successive crossings (transits) of the bright star Sirius. It was a highly consistent interval of 23 hours, 56 minutes and 4 seconds – one sidereal day – that enabled him to check the accuracy of the Tompion clocks.

After months of hard work, Flamsteed did indeed prove the Earth's isochronal rotation, although three centuries later the Observatory's Shortt free-pendulum clock system (90) would prove otherwise. But for Flamsteed, Tompion's specially designed clocks had demonstrated their worth, with extra-long 13ft (4m) pendulums for accuracy that beat every 2 seconds. They also featured an innovative escapement design (see *How does a mechanical clock work?*) to reduce friction and heavy driving weights to keep them going for a year, reducing the need for winding and disruptive maintenance. Yet despite Tompion's best efforts, the clocks were still hampered by layers of dust, a lack of temperature compensation and buffeting winds against the hilltop building. After Flamsteed's death, the two clocks were sold by his widow Margaret and adapted for domestic use, as seen today. They were acquired by the British Museum in 1928 and the National Maritime Museum in 1995.

◉ (and details)
Year-going clock movement, Thomas Tompion, 1676 (ZAA0885)

# FLAMSTEED'S 7FT EQUATORIAL SEXTANT

This view of Flamsteed's 7ft (2.1m) equatorial sextant provide us with a tantalising glimpse of an object that no longer exists. When Flamsteed died in December 1719, his widow Margaret cleared the Observatory of its furniture, instruments and clocks, on the basis that they belonged to Flamsteed himself. This contentious question around the ownership of both property and astronomical data would haunt the fledgling Observatory for the next century, most notably in the legal wrangles over the data collected by the third Astronomer Royal, James Bradley (31).

With no surviving instruments from Flamsteed's time, our knowledge is limited to a set of 12 prints by Francis Place that captured the site in fine detail. Here we can see the equatorial instrument installed within the Sextant House, with the sliding roof and Octagon Room (2) clearly visible in the background. Regrettably, Wren and Hooke's decision to repurpose the site's existing castle foundations had created an observatory that was skewed 13.5° from the local meridian, making it unsuitable for positional astronomy. In response, Wren and Flamsteed designed the Sextant House and adjacent Quadrant House to accommodate both the wall-mounted (mural) and moveable (equatorial) instruments that would become the working heart of the Observatory for the next 40 years.

The Astronomer Royal designed and made this iron-framed equatorial sextant himself, with additional input from Thomas Tompion (3) and blacksmiths at the Tower of London. Unlike earlier versions created by others, Flamsteed's design was more accurate with telescopic sights to see fainter stars, carefully divided scales for greater accuracy and an equatorial mount to make it easier to follow the stars. Working with two assistants, Flamsteed used the instrument to make over 20,000 observations of the angles between the planets and certain bright stars until it was superseded by a new 7ft (2m) mural arc in 1689. Although both the sextant instrument and building have since disappeared, we can still appreciate Flamsteed's innovative design thanks to the inclusion of this scene within his star catalogue, the *Historia Coelestis Britannica* (7).

Print showing Flamsteed's 7ft equatorial sextant, Francis Place after Robert Thacker, about 1676
(PBC1336)

## THE ROYAL OBSERVATORY FROM CROOMS HILL

Created around 20 years after the founding of the Observatory, this painting is a fascinating record of Flamsteed's interest in trialling new telescopes. Since the invention of the instrument around 1609, astronomers had tried to eliminate the colour-distorting effects (chromatic aberration) caused by the curvature and chemical composition of the glass lenses. One early approach was to increase the separation between the lenses (focal length), but this generated long and unwieldy telescopes (2). In March 1676, Flamsteed procured an old 80ft (24.4m) mast from the diarist Samuel Pepys in his capacity as Secretary of the Admiralty, to support a 60ft (18.3m) optical tube. Flamsteed used the new telescope to observe Jupiter's moons but he preferred the results seen with other instruments. In July 1690 he reported with alarm how the mast telescope had begun 'to sway to and fro', prompting the installation of some temporary supports until it could be dismantled a few years later. The trend for lengthy telescopes declined after Dollond's 1758 patent for the achromatic doublet lens, composed of a convex crown lens and a concave flint lens that successfully focused light without colour distortion.

On the right, we can see a small red-brick tower topped by a cupola that marks the location of another one of Flamsteed's innovative, but ultimately doomed, telescopes. In 1669 polymath Robert Hooke tried to use a long vertical telescope mounted in his lodgings at Gresham College, London, to detect stellar parallax, the tiny change in the position of stars as seen from different points along the Earth's orbit. Hooke's results were inconclusive and so Flamsteed installed his own instrument using a 100ft (30.5m) well shaft just beyond the Observatory's boundary. A spiral staircase gave him access to the telescope's eyepiece at the base while the surrounding earth kept the instrument steady. But the damp conditions and poor-quality lens hampered Flamsteed's efforts and he abandoned the instrument after only two recorded observations in 1679. Decades later, his successor James Bradley sought to detect parallax with a similar instrument (15) but instead made a different discovery of the distortion (aberration) of starlight as a consequence of the Earth's motion through space.

*The Royal Observatory from Crooms Hill*, unknown artist, English School, about 1696 (BHC1812)

## 6
## 1714 LONGITUDE ACT

By 1714, Flamsteed had been in post for nearly 40 years but had yet to publish any of his work that could help mariners determine their longitude at sea. A year before, the mathematicians William Whiston and Humphrey Ditton had proposed their idea of having barges loaded with fireworks positioned at regular intervals of longitude. At set times, the crews would fire a shell into the air, thus providing a visual and audible time signal for vessels nearby. With the support of sea captains and merchants, Whiston and Ditton petitioned Parliament for more action in finding longitude solutions, which prompted the convening of a House of Commons Committee on 11 June 1714 to discuss the matter. As President of the Royal Society, Isaac Newton was asked to provide evidence, for which he duly summarised the challenges of the most likely longitude solutions (see *The quest for longitude*), noting that Whiston and Ditton's proposal was better for checking one's longitude, rather than measuring it directly.

In response, the Committee recommended the creation of 'An Act for Providing a Publick Reward for such Person or Persons as shall Discover the Longitude at Sea', which successfully passed through both Houses and received Royal Assent on 20 July 1714, just 12 days before Queen Anne's death. Written on parchment made from goatskin, the original copy of the Act remains in the National Archives today.

Although it was created decades after the Observatory's foundation, the Act and its later revisions would shape the institution's work for the next century. It offered a sliding scale of rewards that were graded according to a project's feasibility in being 'Practicable and Useful at Sea', with a maximum reward of £20,000 – worth millions today – for any scheme that was accurate to within 0.5° longitude, equivalent to about 35 miles (56km) at the equator. Each proposal was to be assessed by a group of Commissioners – later known as the Board of Longitude – drawn from the Admiralty, Royal Society, Board of Trade and others, including the Astronomer Royal. Despite the initial impetus, the Commissioners seemingly only assembled for the first time 23 years later, when John Harrison presented his first marine timekeeper, 'H1' (13), in June 1737.

Longitude Act,
1714 (12 Anne. c.15.)
Parliamentary Archives,
National Archives

**Whereas** it is well known by all that are acquainted with the Art of Navigation That nothing is so much wanted and desired at Sea as the discovery of the Longitude for the safety and quickness of Voyages the preservation of Ships and the lives of men And **Whereas** in the Judgment of able Mathematicians and Navigators severall methods have already been discovered true in theory though very difficult in practice some of which (there is reason to expect) may be capable of Improvement some already discovered may be proposed to the publick and others may be Invented hereafter **And Whereas** such a discovery would be of particular advantage to the Trade of Great Britain and very much for the Honour of this Kingdome But besides the great difficulty of the thing it self partly for the want of some publick reward to be settled as an Encouragement for so usefull and beneficial a work and partly for want of money for tryalls and Experiments necessary thereunto no such Inventions or proposalls hitherto made have been brought to perfection **Be it therefore Enacted** by the Queens most Excellent Majesty by and with the advice and consent of the Lords Spirituall and Temporall and Commons in Parliament Assembled and by the authority of the same That the Lord High Admirall of Great Britain or the first Commissioner of the Admiralty the Speaker of the Honourable House of Commons the first Commissioner of the Navy the first Commissioner of Trade the Admiralls of the Red White and Blew Squadrons the Master of the Trinity House the President of the Royall Society the Royall Astronomer of Greenwich the Savilian Lucasian and Plumian Professors of the Mathematicks in Oxford and Cambridge all for the time being The Honourable Sir Thomas Hanmer Baronet Speaker of the Honourable House of Commons the Honourable Francis Robarts Esquire James Stanhope Esquire William Clayton Esquire and William Lowndes Esquire

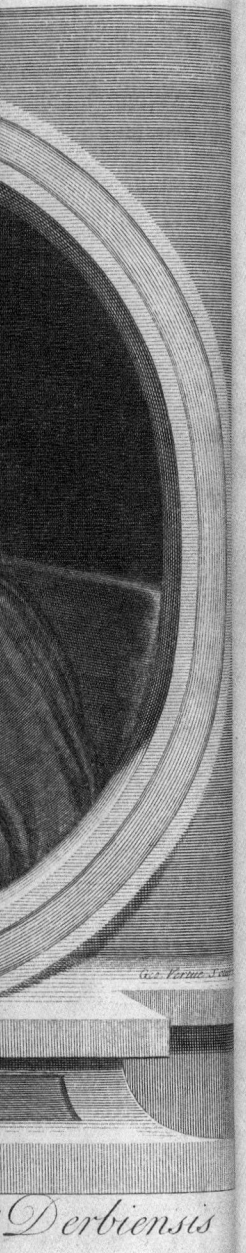

*Derbiensis*
Anno Ætatis 74 Ob[...]
Decem: 31 17[...]

# HISTORIÆ COELESTIS BRITANNICÆ
## VOLUMEN PRIMUM.

Complectens
### STELLARUM FIXARUM
Nec non
### PLANETARUM OMNIUM OBSERVATIONES
*Sextante, Micrometro, &c. peractas.*

Quibus subjuncta sunt
### PLANETARUM LOCA
ab iisdem OBSERVATIONIBUS deducta.

Observante *JOANNE FLAMSTEEDIO*, A.R.
In OBSERVATORIO Regio
### GRENOVICENSI
CONTINUA SERIE
Ab Anno 1675, ad Annum 1689.

LONDINI: Typis H. MEERE. M.DCC.XXV.

## HISTORIA COELESTIS BRITANNICA (1725)

In his revolutionary book *Principia* (1687), Isaac Newton struggled to apply his universal law of gravitation to the Moon's erratic orbit, complaining that 'his head had never ached but when he was studying that subject'. In September 1694, he began to consider plans for a second edition and realised that he needed more lunar data from Flamsteed. It was the start of a decades-long battle between the two scholars as they exchanged angry letters about the quantity, format and availability of data from the Observatory. After initial attempts by the Royal Society to publish Flamsteed's data in 1707, the argumentative Astronomer Royal was excluded from the process and Newton enlisted Edmond Halley to print the material as the *Historia Coelestis Libri Duo* in 1712.

Flamsteed was understandably furious, describing the premature publication of his work as being 'corrupted and spoyled by Dr Halley'. Four years later, a change of government enabled him to use his courtly supporters to recall 300 unsold copies of Halley's 400-copy print run. He extracted the pages that he had already approved in 1707–97 sheets from each book – before burning the remaining pages in April 1716, declaring, 'I made a sacrifice of them to Heavenly Truth.'

Flamsteed only managed to prepare the first two volumes of his three-part catalogue before his death in December 1719, leaving his widow Margaret to consolidate his legacy. She worked with Flamsteed's assistants James Hodgson, Joseph Crosthwaite and Abraham Sharp to compile the final catalogue of 2,935 star positions, which was eventually published in 1725. Written in Latin to make it accessible to scholars across Europe, the catalogue encompassed Flamsteed's entire observational career from Derby to Greenwich and included the rescued pages from the 1712 version. As the first star catalogue based on telescopic observations, it was impressive and renowned for its accuracy. Sadly, Flamsteed's work was rather academic, bulky and expensive, making it more appropriate for other astronomers than mariners at sea for 'perfecting the art of navigation' (see *The quest for longitude*). It would take another four decades before mariners could rely on the Observatory's slimline, affordable and instructive *Nautical Almanac* (19).

*Historia Coelestis Britannica* ('British Star Catalogue'), John Flamsteed, 1725 (PBC1336)

## 8
## *ATLAS COELESTIS* (1729)

Alongside the compilation of her husband's star catalogue, Margaret Flamsteed worked on the production of a star atlas that was eventually published in 1729. Unlike the best-known atlas of the day, Johann Bayer's *Uranometria* (1603), Flamsteed's version was to include several innovative approaches. Firstly, he plotted the stars within an equatorial coordinate system that matched the growing use of these new telescope mounts by astronomers. Secondly, he devised a new type of projection that was intended to reduce distortion when converting the celestial sphere onto a flat page (it was only partly successful). Finally, he wanted to show the constellations as seen by an observer standing on the Earth's surface, rather than employing the traditional perspective of looking down from above, like a celestial globe.

By the time of his death, Flamsteed had only managed to complete the constellation of Orion. Margaret enlisted his former assistants to help her bring the atlas to fruition, with Abraham Sharp painstakingly plotting each star position while Joseph Crosthwaite organised the engraving and printing. The constellation figures were drawn by the artist James Thornhill, who was busy working nearby at the Painted Hall within Greenwich Hospital. Margaret promised to pay the assistants upon completion, but she died soon afterwards in 1730 and left her estate to her great-nephew, leading Crosthwaite to complain bitterly to Sharp: 'What has induced her to act so dishonestly [...] I cannot apprehend.'

Despite its unwieldy size and hefty price tag, the *Atlas* was well received by Flamsteed's peers and underwent two reprints in 1753 and 1781. As astronomers examined the *Atlas* in detail, they realised that Flamsteed had inadvertently recorded the changing position of Uranus several decades before William Herschel discovered the planet with a telescope in March 1781 (30). The Astronomer Royal had recorded the 'star' on least seven occasions during the period 1690–1715 and initially gave it the designation 34 Tauri, seemingly unaware that successive observations of other 'stars' were in fact our first recorded glimpses of the elusive seventh planet.

◐ (and overleaf)
Constellations of Orion, Taurus, Cassiopeia, Cepheus, Ursa Minor and Draco, *Atlas Coelestis* ('Heavenly Atlas', 2nd edition), John Flamsteed, 1753 (PBC1346)

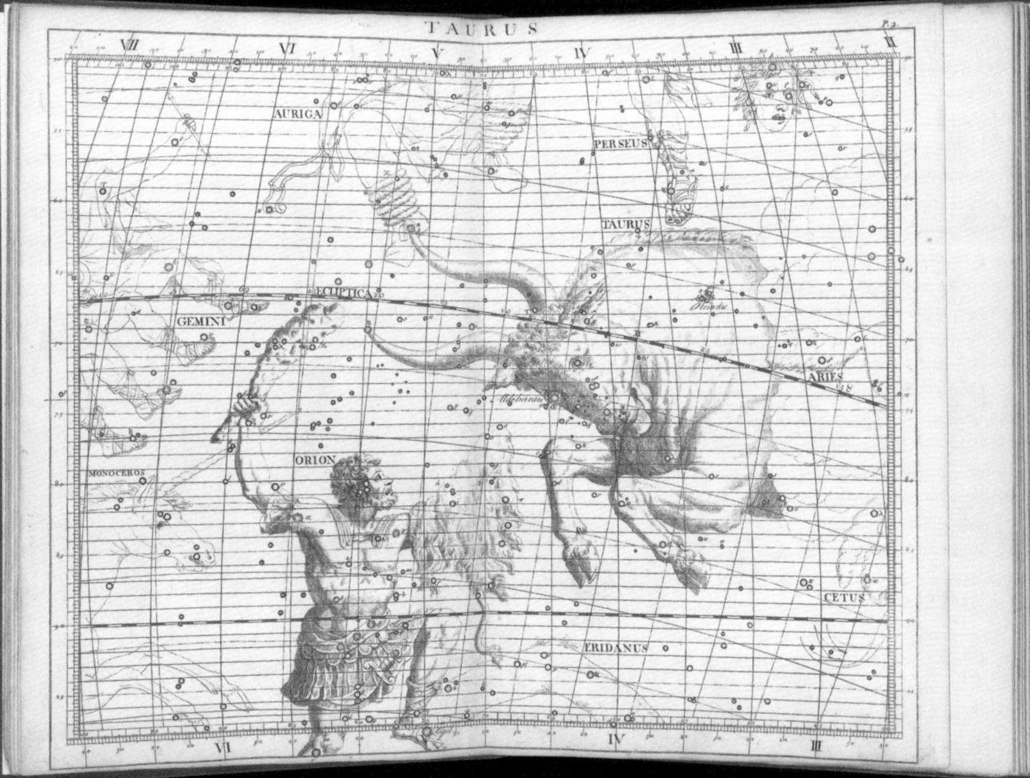
TAURUS
VII
VI
V
IV
III
II
AURIGA
PERSEUS
TAURUS
ECLIPTICA
GEMINI
ARIES
Aldebaran
ORION
MONOCEROS
CETUS
ERIDANUS
VI
IV
III

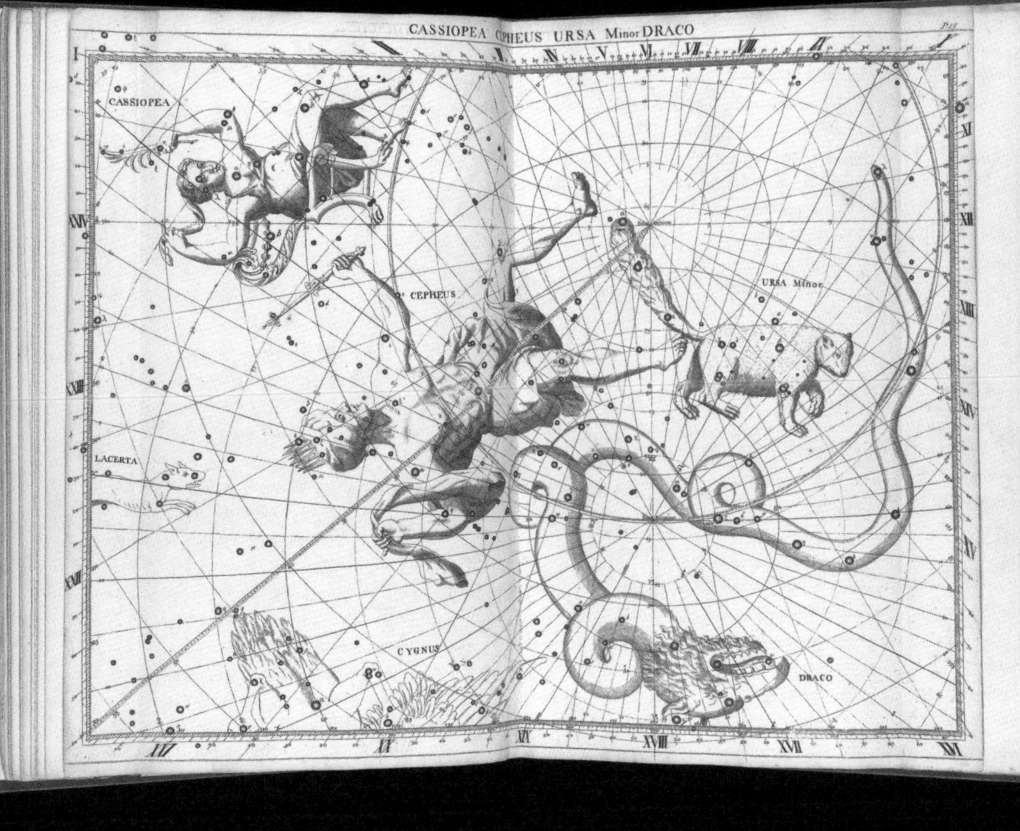
CASSIOPEA CEPHEUS URSA Minor DRACO
I
CASSIOPEA
CEPHEUS
URSA Minor
LACERTA
CYGNUS
DRACO
XI
XII
XIII
XIV
XV
XVI
XVII
XVIII
XIX
XX
XXI
XXII
XXIII
XXIV

# Edmond Halley

## 1656–1742

Edmond Halley's tenure at Greenwich was the final chapter in a very productive and eventful life. Unlike many of his fellow Astronomers Royal, Halley was famous and well-respected long before his arrival at the Observatory. Similarly, his life has since been defined by his comet fame, rather than his achievements at Greenwich. Born into an affluent London family, Halley began making astronomical and magnetic measurements as a teenager before continuing his studies at Oxford. Like many of his peers, he quit before completing his degree and embarked on a life of travel and scientific exploration during which he rubbed shoulders with both royalty and the scientific elite. These experiences gave him the opportunity to observe and interpret global phenomena, leading him to create the first charts of magnetic variation, trade winds, and the stars visible from the southern hemisphere.

In 1682 he married Mary Tooke and settled into the London scientific community through his involvement with the Royal Society. With his own interests in planetary and cometary orbits, Halley recognised the significance of Newton's groundbreaking mathematical studies and encouraged the reclusive scholar to publish his famous book, *Principia* (1687). He also recognised the implications of other people's ideas to make predictions that were verified decades after his death, most notably during the transit of Venus expeditions of the 1760s, and in the return of the comet that bears his name. In addition to astronomy, Halley applied his skills to calculating life expectancy tables for insurance companies, designing a diving bell for extracting goods from shipwrecks and mapping magnetic variation across the Atlantic as a means of measuring longitude.

Upon his arrival at Greenwich in March 1720, Halley started afresh with the purchase of new instruments and timekeepers, including a transit telescope that he used to accurately define the first Greenwich meridian to appear on a map. Now aged in his mid-60s, the second Astronomer Royal settled into a programme of lunar observations to learn more about the Moon's orbit. Regrettably, the poorly maintained instruments and unreliable results have since been criticised by Halley's successors and historians alike, creating a rather sad ending for such a distinguished and visionary thinker.

*Edmond Halley*, Godfrey Kneller, before 1721
(BHC2734)

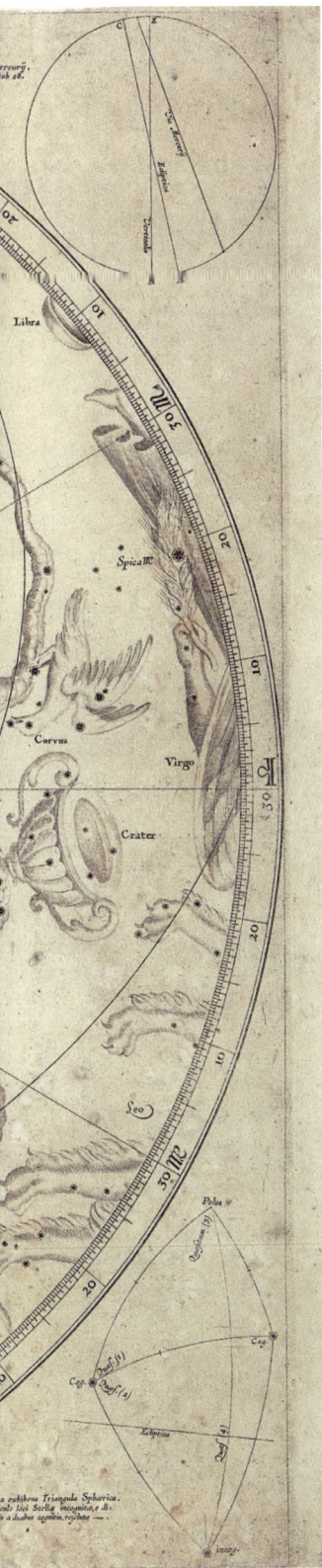

# CHART OF THE SOUTHERN CELESTIAL HEMISPHERE

Published just two months after Halley's return from Saint Helena in 1678, this star chart encapsulates many of his interests that would subsequently define his career and pave the way for his appointment as Astronomer Royal. During his undergraduate studies at Oxford, Halley had realised that Flamsteed's star catalogue (7) would only include stars visible from the northern hemisphere. He boldly approached the Royal Society with an ambitious plan to travel on an East India Company ship to the remote island of Saint Helena, promising both to map the southern skies and to witness the forthcoming transit of Mercury in October 1677 as a means of measuring the distance to the Sun (see *Transit of Venus expeditions*).

With a letter of support from Charles II and a suite of telescopes, clocks and quadrants paid for by his wealthy father, Halley undertook the three-month voyage in 1677. He set up his observatory on the island's central peak now named 'Halley's Mount' and endeavoured to map as many stars as possible, despite the frequent cloud cover.

After a year of observing, Halley returned to London to publish this celestial chart, the first record of the southern hemisphere stars to be compiled via telescopic observation, followed by his catalogue of 341 stars, the *Catalogus Stellarum Australium* (1679). Halley included 12 new southern constellations that had previously been devised by Dutch navigators in the 1590s and he also inserted his own creation. With a nod to his royal sponsor, he created 'Robur Carolinum' (Charles's Oak) to commemorate the oak tree in which the King had hidden from Parliamentarian soldiers during the English Civil War in 1651 (see centre section, below the centaur's hooves). The surrounding diagrams about the transit of Mercury (upper corners) and calculating stellar coordinates (lower corners) were included as impressive examples of the author's mathematical expertise.

Halley's youthful confidence and self-promotion paid off: on the basis of this catalogue and its charts, he was proposed by Robert Hooke (5) as Fellow of the Royal Society and thus secured his place in the history of science at just 22 years old.

←
Chart of the southern
celestial hemisphere,
Edmond Halley and
James Clerk, 1678
(G200:3/2)

# HALLEY'S TRANSIT TELESCOPE

When Halley arrived at the Observatory in March 1720, he set about replacing the clocks and telescopes that had been removed by Margaret Flamsteed. He secured £500 from the Board of Ordnance and used £30 to purchase a telescope that was 5ft 6in. (1.7m) long and 1.75in. (4cm) in aperture, reputedly made by Robert Hooke. He then used another £30 to construct a building in the gap between the western summerhouse and the wall of Flamsteed House. Here, he installed both a 4ft (1.2m) stone pier to support the new telescope and a week-duration clock by George Graham to time the observations. He commenced work on 1 October 1721 and the telescope remained in periodic use until 1750.

While Flamsteed had indeed aligned his instruments with the local meridian, Halley wanted to go a step further by accurately defining a meridian according to the stars. Unlike a fixed mural quadrant that could only be used facing south, Halley's transit telescope could be wall-mounted at each end of its axis and then rotated vertically in an arc from south to north. This meant that he could time the transit of circumpolar stars twice every sidereal day: once on the southern part of the meridian and then again 12 hours later, when they crossed the northern part of the meridian. Halley knew that the instrument was correctly aligned with the true meridian when the time interval between north and south transits was constant. He then installed a south mark on one of the walls in the park as a convenient check for any alignment errors. For the telescope's mount, Halley opted for the unusual off-centre style devised in 1689 by the Danish astronomer Ole Rømer but the reasons for his choice are unclear.

At the time of the instrument's construction, English cartographers relied on a meridian passing through St Paul's Cathedral as 0° longitude but, within 20 years, people had started to use Halley's new Greenwich meridian instead, as shown by the chart *A Description of the Sea Coast of England and Wales* published in 1738 by Samuel Fearon and John Eyes.

Ole Rømer at his transit telescope, Peder Horrebow, *Basis astronomiae*, 1735.
Linda Hall Library of Science, Engineering and Technology

Halley's transit telescope, attributed to Robert Hooke, 1721 (AST0979)

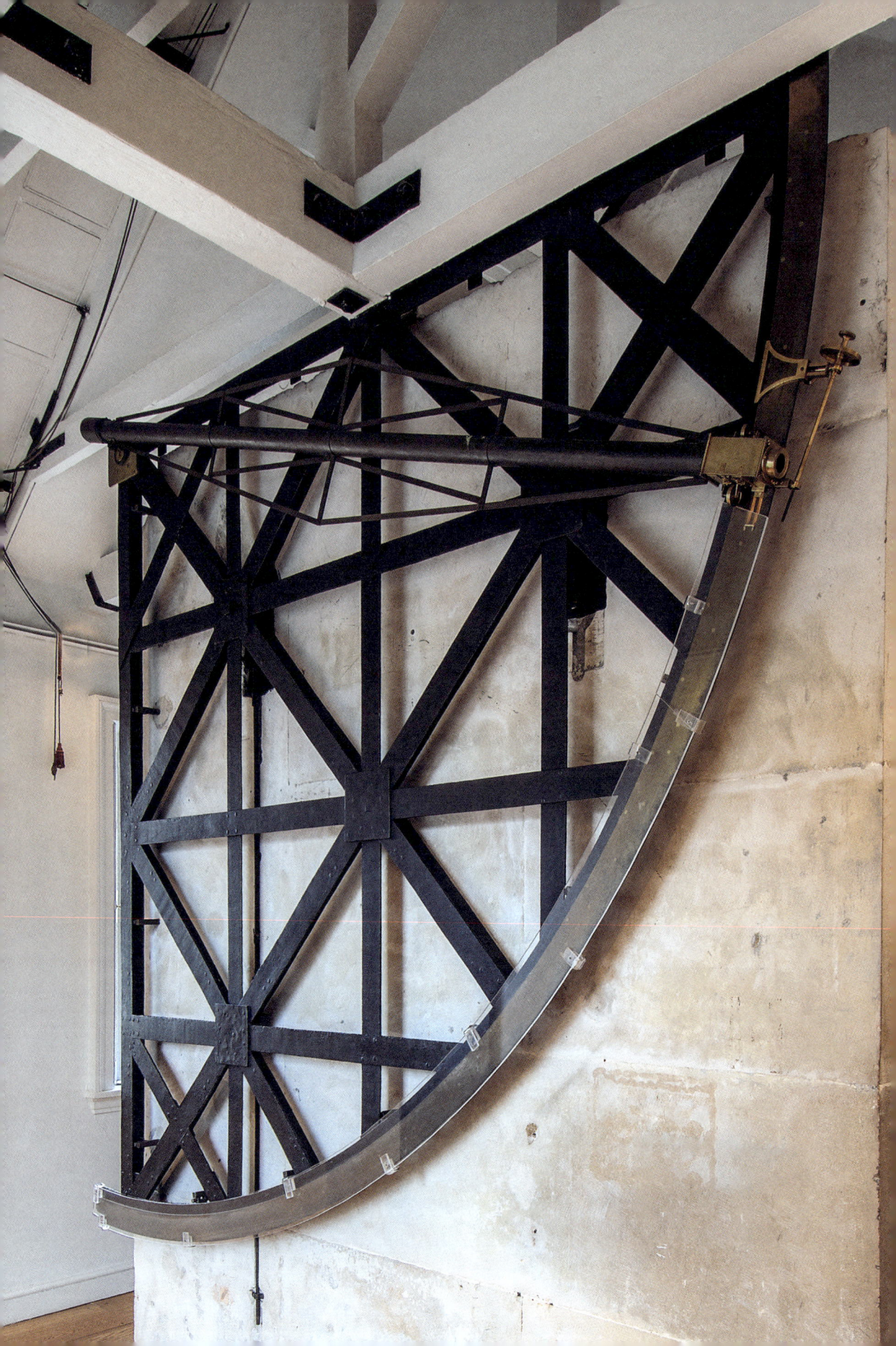

## 11

## HALLEY'S 8FT MURAL QUADRANT

Having established a meridian for accurate timing with a transit telescope (10), Halley now sought to build upon his measurements of the Moon and planets. He commissioned an 8ft (2.4m) radius mural (wall-based) quadrant that was installed on a new stone pier, just a few metres away from Flamsteed's meridian wall, which was now subsiding. The instrument makers George Graham and Jonathan Sisson created an instrument that later became the prototype for similar ones at observatories across Europe. Two finely divided scales, an inner one (0–90°) and an outer one (divided into 96 parts), read with a vernier scale, enabled the astronomer to cross-check the readings for greater accuracy, while the instrument's iron frame helped keep it rigid. Graham also supplied a month-going regulator, 'Graham 2' (12), that provided sidereal time.

Halley installed the mural quadrant facing south on the eastern side of the stone pier, with the intention of purchasing another to face north but the funds never materialised. Whenever the skies were clear, Halley opened up the roof hatches to view the stars and planets crossing the instrument's meridian. Having positioned the telescope towards the relevant part of the sky, he centred the approaching object within the eyepiece and read off the zenith distance (the angle between the star and the north celestial pole) on the scales as a measure of its declination. It was tedious work in freezing cold conditions and, as far as we know, Halley worked alone with no assistant, even continuing to observe after suffering paralysis in his right hand in 1737.

Following Halley's death in January 1741, the third Astronomer Royal, James Bradley, arrived at Greenwich to find the instrument in a poor state of repair: the counterpoise that kept the instrument on balance (now missing) was scraping along the sagging timbers of the ceiling while the iron frame had warped under its own weight and was no longer correctly aligned. Bradley asked Graham and Sisson to make the necessary adjustments and the instrument continued to be used by the third Astronomer Royal and his successors until 1812.

←
Halley's 8ft mural quadrant, George Graham and Jonathan Sisson, 1725
(AST0970)

## How does a mechanical clock work?

We have all seen different clocks and watches, but what is really happening inside? The Observatory is home to a wide range of mechanical timekeepers, all of which have these essential components:

- **movement:** The name for the entire mechanism of the clock that keeps regular time intervals.
- **energy source:** Power provided by pendulum weights (gravity) or coiled springs (potential energy).
- **wheels:** Interlocking gears (wheel train) that transfer energy from the source to other sections of the clock, making parts move at different speeds to measure hours, minutes and seconds.
- **oscillator:** The component that swings or oscillates at regular intervals, such as the swing of a pendulum or the rotation back and forth of a weighted wheel around a spiral spring (balance wheel).
- **escapement:** The mechanism that keeps the oscillator running and controls the speed at which the energy is released from the power source into the wheels, making the distinctive 'tick tock' sound and comparable to a valve, clutch, or even a turnstile.
- **indicator:** The visible part that shows the time, either as a dial, hands, moving figurines or even as the sound of a bell.

### What does 'regulator' mean?

Astronomers relied on highly accurate pendulum clocks called 'regulators' to provide a reference when checking other clocks. To ensure reliability, clockmakers added special features to regulators to compensate for temperature variations that might change the length of the pendulum and cause the timekeeper to speed up or slow down.

One solution, as devised by George Graham in 1721, was to create a brass pendulum with a jar of liquid mercury as the 'bob'. As the metal pendulum rod expanded in warmer temperatures, the mercury also expanded within the jar. This expansion shifted the centre of mass upwards, causing the pendulum to swing with the same duration (period) as before. The opposite effect occurred in cold temperatures. Another option, as devised by John Harrison in 1726, was to create a pendulum from parallel alternate rods of brass and steel (gridiron) that expanded and contracted at complementary but opposite rates, keeping the effective length of the pendulum constant.

For maximum clarity when checking the time, regulators were also fitted with separate dials for each time interval (hours, minutes, seconds), rather than a combined dial. They were usually set to sidereal time, as measured by the stars, which was then converted into mean solar time for use on civilian clocks and watches. Regulators were sometimes described by the amount of time between winding, such as 'eight-day' or even 'month-going'.

### How good is my clock?

Various standards are used to compare timekeepers. A clock's 'rate' describes how much it will speed up or slow down and is usually stated as a gain (+) or loss (-) in seconds per day. A clock's 'error' is its cumulative rate over time. 'Accuracy' is a comparison between the time shown on a clock – corrected for its error – against a known reference time, such as an observatory clock or atomic clock. 'Precision' is a measure of consistency and stability. A highly precise clock will have a small rate that varies little over time.

↑
Regulator with minutes (outer scale), seconds (upper centre dial) and hours (lower centre dial), William Hardy, about 1825 (ZAA0606)

## REGULATOR NO. 1 BY GRAHAM

Impressed by the success of Graham's week-going clock alongside the transit telescope (10), Halley ordered another two accurate observatory clocks (regulators) from the Fleet Street clockmaker in 1725, one for the Quadrant Room and one for the Great Room (Octagon Room, 2). These month-going regulators, housed in mahogany longcases, were supplied with one of the best horological innovations of the day, namely a 'dead-beat' escapement (see *How does a mechanical clock work?*) that reduced friction and improved accuracy. Graham based his work on a design that had been proposed by his mentors, Richard Towneley and Thomas Tompion, over 40 years earlier. The two Observatory regulators, later designated 'Graham 1' and 'Graham 2', were originally supplied with simple pendulums but were later modified with improved features and remained in use for over 250 years in various locations, both at Greenwich and Herstmonceux (see *Epilogue*).

Halley called upon Graham's expertise a few years later when a carpenter from Barrow upon Humber, John Harrison, arrived at the Observatory with his idea for a timekeeper that could be used to determine longitude at sea. Halley could offer little technical advice and referred Harrison to Graham instead. After a lengthy discussion lasting ten hours at the clockmaker's home, Harrison convinced him of his ideas and Graham went on to become a pivotal figure in the Lincolnshire man's career, both as a conduit into the London clockmaking community and even as a financial investor in the first marine timekeeper, 'H1' (13).

Graham's association with the Observatory dwindled during the 1730s as Halley became increasingly frail and unable to secure funding for new instruments. But the clockmaker faced no shortage of work with additional orders from the French Academy of Sciences and the wealthy amateur astronomer James Bradley, who would succeed Halley as Astronomer Royal in 1742. A keen amateur astronomer himself, Graham also developed a geared model of the Sun, Moon and Earth for Charles Boyle, the fourth Earl of Orrery, whose name became synonymous with these devices (43).

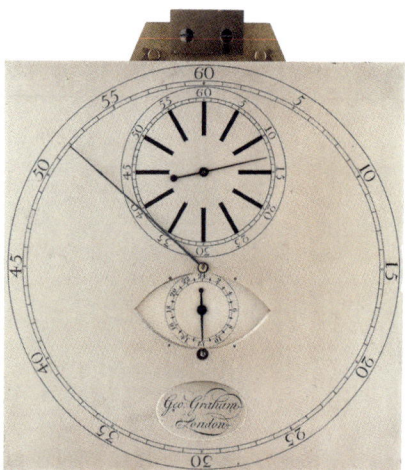

➲ (and detail)
Regulator no.1, George Graham, 1725 (ZBA2211)

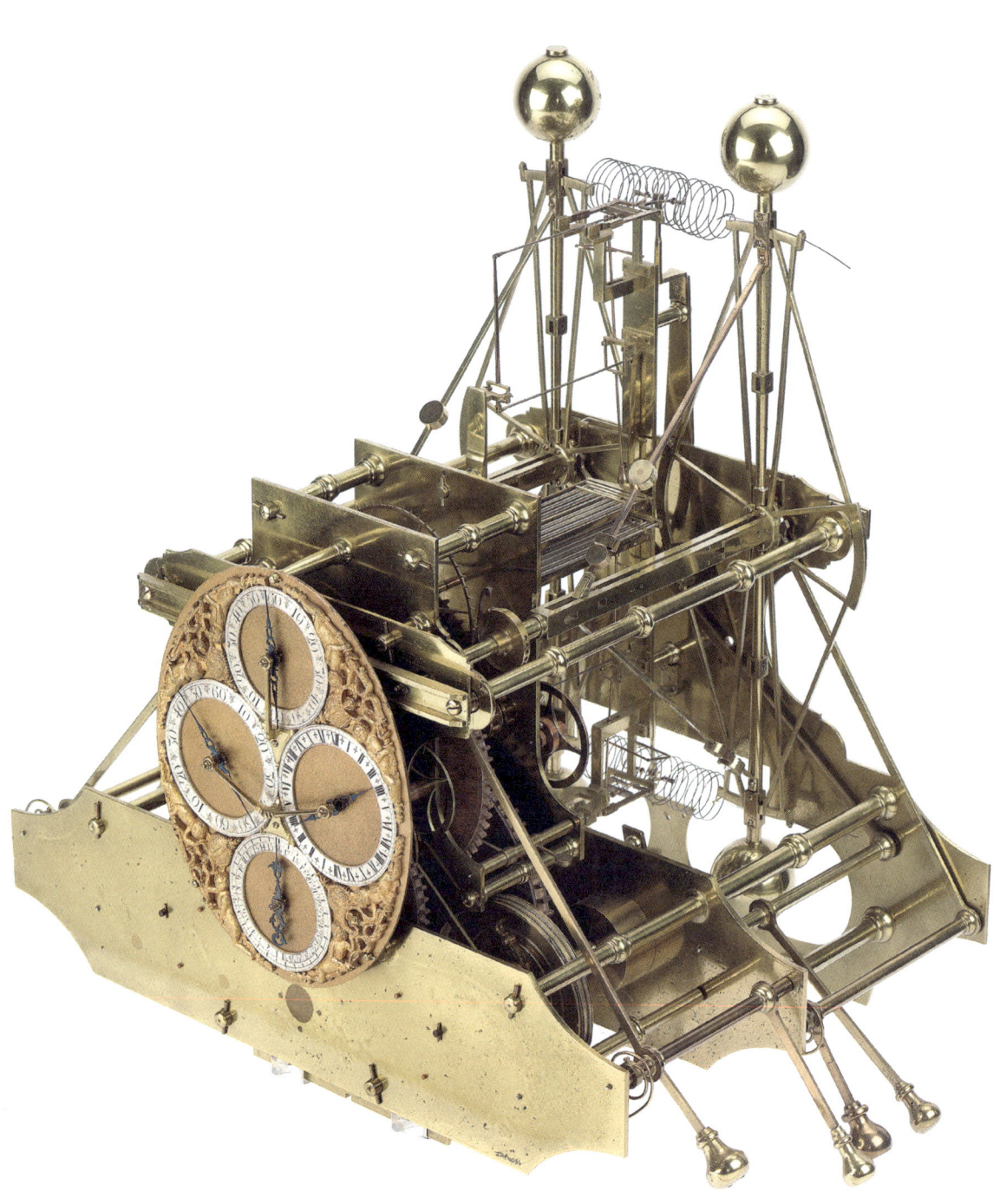

## 13

## MARINE TIMEKEEPER BY HARRISON, 'H1'

With George Graham's financial support and introduction to the London clockmaking community (12), John Harrison set to work on his first marine timekeeper for keeping a reference time at sea (see *The quest for longitude*). Standing 25in. (63cm) high and weighing 75lb (34kg), the timekeeper has three main features to address the challenge of keeping accurate time on oceanic voyages. Firstly, Harrison replaced the usual swinging pendulum with a spring-bar balance system that was less affected by gravity and the motion of the ship. Secondly, he took inspiration from his 1726 invention of the gridiron pendulum to add a series of steel and brass rods to the balances. These parts expanded and contracted at different rates, effectively cancelling out any temperature variation that might cause the timekeeper to speed up or slow down. Finally, he used his experience as a carpenter to select the best timbers for the wooden components. The main wheels were made of oak but for the smaller components he chose lignum vitae, a tropical hardwood that secretes its own natural oil. This reduced energy losses through friction and negated the need for oiling on long voyages. Harrison also reduced frictional losses by including his earlier invention of the grasshopper escapement, whose motion resembled the jumping hind legs of the insect. In addition, he used bronze alloys for additional strength and minimised the use of steel components that could rust at sea and be affected by magnetism.

After five years' work with his brother James in Barrow upon Humber, Harrison was ready to submit his timekeeper for testing. The duo had already undertaken some preliminary trials on a river barge but, with support from the Royal Society and the Admiralty, Harrison took his timekeeper to sea on board the warship *Centurion* in May 1736. They headed towards Lisbon under the watchful eye of Captain Proctor who was instructed 'how it [the timekeeper] will succeed at Sea, you will partly be a Judge'. It was a long and challenging week for Harrison as the machine performed badly and he struggled with constant seasickness. The return voyage on the *Orford* was much more successful, with Harrison even using the timekeeper to alert the crew that the ship was approaching the rocky coast of Penzance, rather than Dartmouth, which was still over 60 miles (96km) away. Captain Proctor later praised Harrison for averting shipwreck.

←
Marine timekeeper 'H1',
John Harrison, 1735
(ZAA0034)

# 14

## MARINE TIMEKEEPER BY HARRISON, 'H2'

At a meeting held on 30 June 1737, the Board of Longitude (6) appraised Harrison's first marine timekeeper, 'H1' (13) and decided to award him £250 to work on an improved design, followed by another £250 once the machine had been tested on a voyage to the West Indies. Harrison had moved to London the year before with his second wife, Elizabeth, and their sons, John and William. The family initially settled in Leather Lane, Holborn, but in 1739 moved the short distance to Red Lion Square.

Harrison was now ideally placed to seek specialist input from the city's metalworkers with expertise in making springs, engraving and polishing. He continued to work mainly with brass, apart from a few key components made from the tropical hardwood lignum vitae, similar to his work on 'H1'. Harrison's main innovation was to add an extra spring system, known as a 'remontoire', which rewinds every 3 minutes 45 seconds to reduce any variation in power delivered to the escapement that might cause the timekeeper to speed up or slow down.

The timekeeper was completed in under two years and was originally positioned within a gimballed mount, now missing. Similar but more rectangular than Harrison's first design, this second version, later known as 'H2', stands 26in. (66cm) high and weighs slightly more at 86lb (39kg). Unfortunately, Harrison realised at a late stage that the bar balances were more susceptible to motion than expected and so he decided to start afresh with a new timekeeper that was sustained by a wheel balance. To raise the necessary funds, he submitted a certificate from his mentor George Graham asserting the 'usefulness' of his machines to the Board of Longitude, which duly awarded him £500 at its meeting on 16 January 1741. The funding was based on the expectation that the third timekeeper would be ready for trial at sea by 1 August 1743 but, in reality, it would take another 20 years and a complete change of approach before Harrison's ideas could be tested on a voyage to the West Indies.

◐ (and detail)
Marine timekeeper 'H2',
John Harrison, 1739
(ZAA0035)

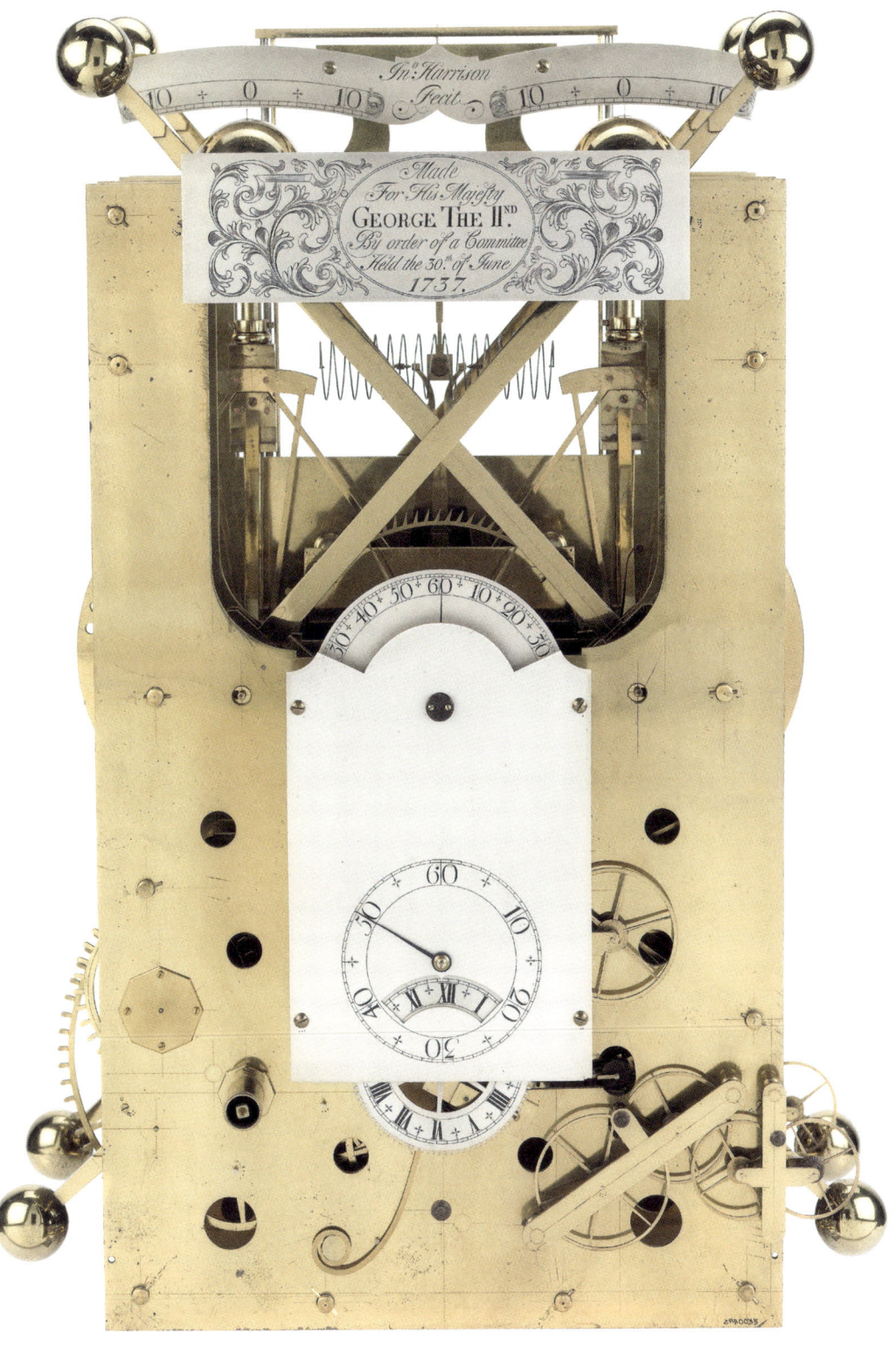

THIRD ASTRONOMER ROYAL, 1742-62

# James Bradley
## 1692–1762

As Halley's health declined, he offered to resign as Astronomer Royal and expressed his preference for James Bradley as successor. The two men had known each other for over 20 years, starting with an introduction by Bradley's uncle, the amateur astronomer Reverend James Pound who had an observatory at Wanstead in north-east London. From 1718 onwards, Halley presented Bradley's work on aurorae, nebulae, double stars and observations of Mars to the Royal Society, thus cementing the young astronomer's reputation as a talented observer.

Halley's rapid decline and eventual death in January 1742 negated the need for his resignation but his wishes were honoured and Bradley began work at Greenwich five months later. Originally from Gloucestershire, Bradley had initially worked as a vicar (hence his clerical attire in this portrait) but he relinquished this career path in 1721 to return to his university as Savilian Professor of Astronomy at Oxford. Apart from some lectures at the Ashmolean Museum, Bradley's academic responsibilities were minimal, allowing him to focus on observations of Jupiter's moons, the rings of Saturn, Mercury, Venus and comets. Reverend Pound died in November 1724 but Bradley continued to use his uncle's observatory for several more years, most notably for his famous discovery of the aberration of starlight.

Upon his arrival at Greenwich, Bradley found the buildings and instruments to be in a poor state: Flamsteed's Sextant and Quadrant Houses had been converted into pigeon lofts, while Halley's mural quadrant was wedged against the roof hatch. The new Astronomer Royal used his contacts at the Royal Society to successfully petition the Board of Ordnance to invest in new buildings and improved instruments.

Two years after his appointment, Bradley married Susannah Peach of Chalford, Gloucestershire, and they welcomed the arrival of a daughter, also called Susannah, in 1746. After the death of Mrs Bradley in 1757, the widowed Astronomer Royal relied more on his wife's family as his own health declined. He died in Chalford on 13 July 1762. His voluminous set of 60,000 observations was bequeathed to his daughter but it would take another 36 years of legal wrangling to secure and publish Bradley's work.

←
*James Bradley*, Benjamin Wilson, about 1750
(ZBA0722)

# BRADLEY'S 12.5FT ZENITH SECTOR

This instrument played a crucial role in helping us understand the Earth's motion through space. In 1725, the wealthy politician Samuel Molyneux invited Bradley to use a new 24.5ft (7.5m) vertical telescope (zenith sector) at his private observatory in Kew. Their goal was to measure the tiny, apparently seasonal shift in the position of the stars (parallax) as evidence of the Earth's orbit around the Sun, as proposed by Nicolaus Copernicus in 1543. To minimise the distorting (refractive) effects of the atmosphere, Bradley and Molyneux decided to observe the star Gamma Draconis, which is visible directly overhead (zenith) as seen from London, through the thinnest part of atmosphere. Just like previous attempts by Robert Hooke and Flamsteed (5), the initial results were disappointing.

Two years later, Molyneux was appointed to the Admiralty and had less time for astronomy, leaving Bradley to continue the project at his uncle's observatory in Wanstead, north-east London. The astronomer commissioned this second, shorter instrument from George Graham (10, 11, 12) that could be swung a few degrees north or south of the zenith to view around 200 nearby stars for comparison. After a year of observations, Bradley had indeed detected a slight shift in the apparent position of the stars but not at the expected time of year – the results were three months out. Having rejected all other possible explanations, Bradley concluded that the observer's view of the starlight was affected by the Earth's actual motion through space, rather than a change in perspective across the seasons. Bradley called this effect 'aberration' and used the results to estimate the time taken for light to travel from the Sun to the Earth as 8 minutes and 13 seconds, which is remarkably close to today's value of 8 minutes and 20 seconds. Buoyed by the instrument's success, he continued to measure Gamma Draconis for another 18 years and detected another effect known as 'nutation', whereby the Moon's gravitational pull causes the Earth to wobble slightly on its axis.

All these insightful results cemented Bradley's reputation as a skilful observer and he later persuaded the Board of Ordnance to purchase this instrument for the Observatory, where it remained in use until 1837.

➲
12.5ft (3.8m) zenith sector,
George Graham, 1727,
with a later eyepiece by
Troughton and Simms
(AST0992)

59

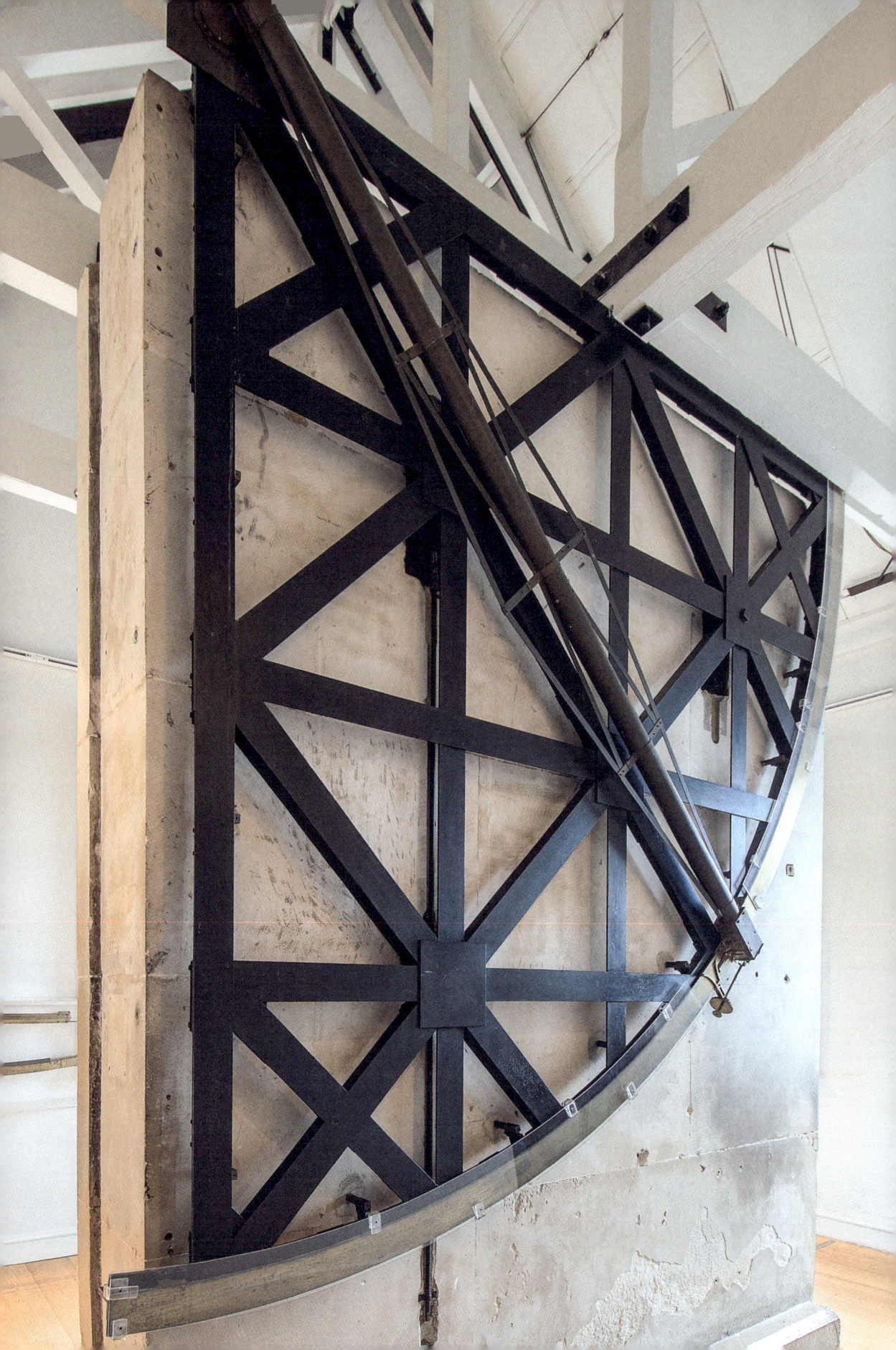

## 16
## BRADLEY'S 8FT MURAL QUADRANT

Upon his arrival at Greenwich in 1742, Bradley reassessed Halley's observations and found them to be defective, possibly because of Halley's failing eyesight and the lack of instrument maintenance during his twilight years. After six years of valiantly trying to use Halley's deteriorating instruments, Bradley successfully petitioned the Board of Ordnance to grant him £1,000 for upgraded versions. By February 1749 he had commissioned John Bird to create a counterpart to Halley's mural quadrant made by Graham and Sisson (11). Bird examined the original quadrant and realised that the instrument had distorted under its own weight, possibly due to connecting parts made of iron rather than brass. In response, he created a second quadrant made entirely of high-quality brass that performed well, with no distortion caused by weight or changes in temperature. The London maker became famous for his accurate instruments in which the scales were carefully divided by hand, leading to commissions for similar instruments from observatories in Oxford, Paris and Saint Petersburg. Nearly 20 years later, Bird received £500 from the Board of Longitude (6) for providing a 'full and complete description' of this method so that others could reproduce his work.

The new 8ft (2.4m) radius quadrant came into use at Greenwich during the summer of 1750 and was initially installed facing north. It became Bradley's workhorse for the next three years as he made observations of Polaris and the circumpolar stars (see *Glossary*) to accurately determine the Observatory's latitude of 51° 28' 39.5" N. Once his observations were completed, Bradley swapped the mural quadrants in 1753 so that he could use the new Bird instrument facing south. Based on his previous work with the zenith sector (15), Bradley also recognised the significance of local atmospheric disturbances that could distort the observations and so he installed a barometer and thermometer nearby.

A few years later, Bradley relied on this instrument to assess new lunar tables submitted to the Board of Longitude by the Hanoverian cartographer and mathematician Tobias Mayer. By comparing the tables against his own observations of the Moon, as seen from Greenwich during the previous five years, Bradley could indeed verify Mayer's work. These tables marked an important milestone in the subsequent development of the *Nautical Almanac* (19).

Mural quadrant, John Bird, 1750 (AST0971)

Portrait of John Bird featuring his design for the mural quadrant, Valentine Green, 1776 (PAF3435)

## 17

## BRADLEY'S TRANSIT TELESCOPE

Unlike Halley's asymmetric transit telescope (10), Bradley's instrument was to be mounted centrally between two stone piers. Flamsteed had tried to measure both star coordinates using a 7ft (2.1m) mural arc (installed in 1689), but 60 years later, Bradley realised that he could use separate instruments for better accuracy, namely a mural quadrant to measure a star's vertical position (declination) and a transit instrument to measure its horizontal position (right ascension). He commissioned this instrument from John Bird (16) in February 1749 and made the first observation on 2 September 1750, diligently watching each bright star as it appeared to cross a series of five wires visible across the aperture, while listening for the pendulum beat of a regulator by Graham (12). Bradley's observing technique became known as the 'eye-ear method' by which he could measure sidereal time to the nearest tenth of a second. The telescope's alignment was checked against meridian markers that were installed 0.5 miles (0.85km) away, north and south. Bradley relied on his nephew John Bradley as observing assistant, entrusting him to care for the Observatory when away in Oxford.

The transit instrument remained the Observatory's main instrument for defining the Greenwich meridian for the next six decades with occasional upgrades such as a new achromatic object-glass by Dollond installed in 1772, and mahogany shields for protecting the observer during solar transits, in 1784. It remained in use until 5 July 1816 when it was dismantled to make way for a newer 10ft (3m) version by Edward Troughton.

Despite its simplistic appearance, this instrument is an important reminder of the Observatory's contribution to navigation both at sea and on land. During Maskelyne's time, it was used by a succession of assistants whose observational data was essential for computing the values of the *Nautical Almanac* (19) that ultimately gave mariners their location at sea, relative to Greenwich. Similarly, in the 1790s, the telescope was used to define the Prime Meridian (0° longitude) for the first Ordnance Survey map, published in 1801.

◉
8ft transit telescope,
John Bird, 1749 (AST0980)

◉
Drawing of Bird's transit instrument in place at Bradley's transit room, John Charnock, about 1790 (PAF2956)

◉
Detail of *General survey of England and Wales: an entirely new and accurate survey of the county of Kent*, Captain William Mudge, Thomas Foot, William Faden, 1801 (PBE7449/1)

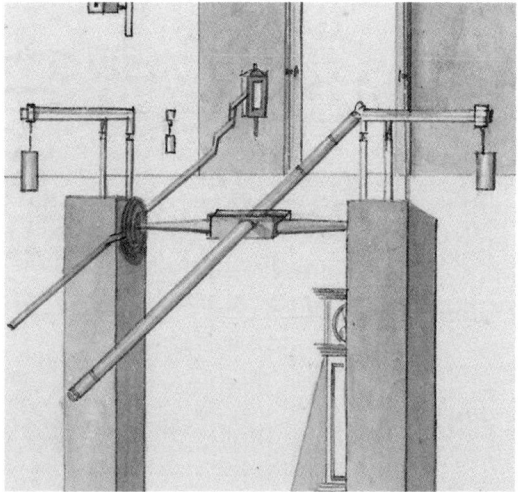

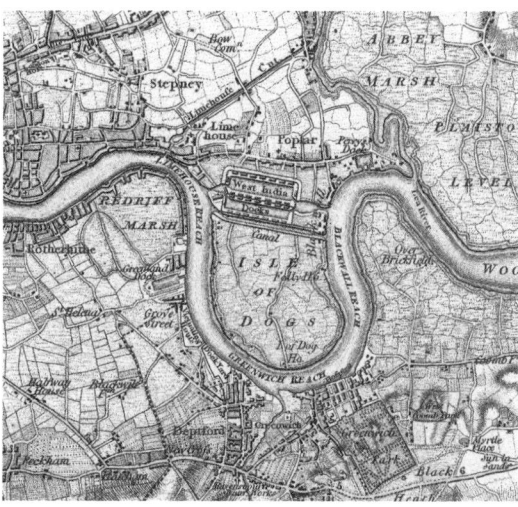

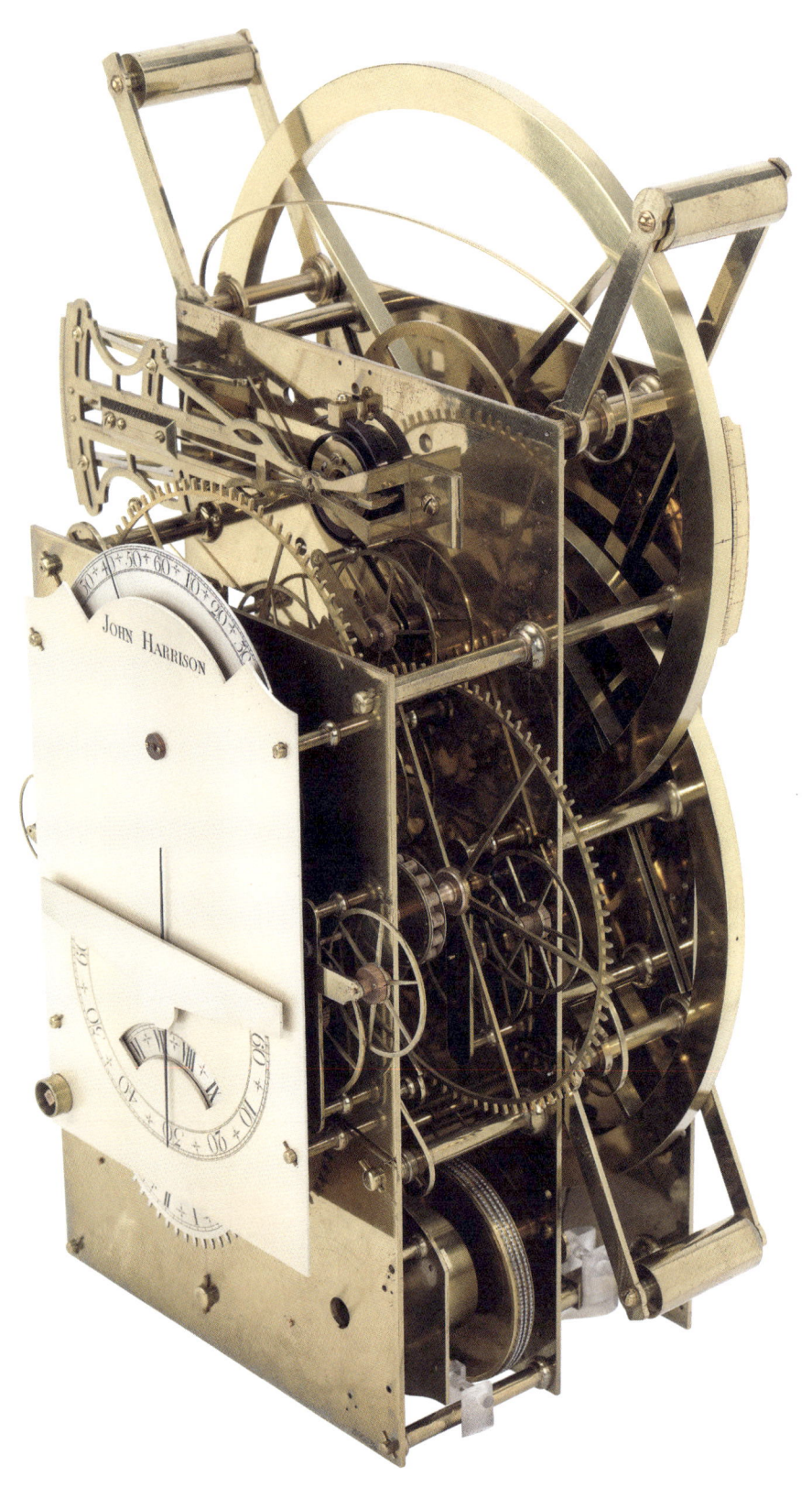

# 18
# MARINE TIMEKEEPER BY HARRISON, 'H3'

In January 1740, Harrison started work on his third marine timekeeper, a task that would occupy him for the next 20 years and would stretch his knowledge of physics, materials and engineering. Harrison's perseverance resulted in three main innovations: the first was the change from dumbbell-type balances to circular wheel balances for greater consistency during motion. The second was his invention of a bimetallic strip composed of flat pieces of brass and steel riveted together and positioned within a violin-shaped brass frame. With one end fixed, the other could bend in response to temperature change and adjust the balance spring accordingly. This principle was used centuries later in domestic technologies such as kettles and toasters. The third innovation was a set of caged roller bearings in which the central pivot was surrounded by a circle of bearings that evenly distributed the forces and minimised any friction on the pivot itself, yet another principle that is still used in industrial machinery today.' Finally, Harrison developed a new remontoire that rewound every 30 seconds to ensure a constant driving force. The whole machine was intended to be housed in a glazed brass frame case suspended within gimbals.

The clockmaker initially presented his 23in. (59cm) tall machine to the Board of Longitude on 4 June 1746, accompanied by a petition from 12 supporters who vouched for the plausibility of his device, even though they admitted that 'it does not go so well at present as he expected'. The Board duly awarded him a further £500, followed by additional payments over the next decade. The machine became a defining part of Harrison's work, earning him the Royal Society's Copley Medal in 1749 and later featured in his portrait by Thomas King in 1767. But, despite 20 years' work, the machine never fulfilled its potential and Harrison abandoned it to progress his fourth design (21). In a curious twist of parallel fates, his third machine would prove to be equally challenging and time-consuming for the horologist Rupert Gould (94) when he restored it over 170 years later.

Marine timekeeper 'H3', John Harrison, about 1759 (ZAA0036)

Details showing H3's violin-shaped frame containing the bimetallic strip for temperature compensation (left) and the caged roller bearing on the reverse (right) (ZAA0036)

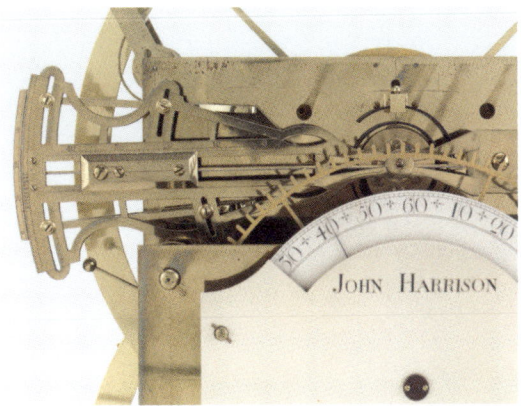

# Nathaniel Bliss

## 1700–64

Nathaniel Bliss was an experienced mathematician and observer whose tenure at Greenwich was cut short by his untimely death at the age of 63. Like his predecessor Bradley, Bliss studied at the University of Oxford and began a clerical career in 1736 as rector of a city church, supported by his wife Elizabeth and their growing family. He attended Bradley's lectures at the Ashmolean Museum and was later appointed as Savilian Professor of Geometry upon Halley's death in 1742. He established himself as a lecturer in arithmetic, algebra, logarithms and trigonometry, working with small groups of students over the course of three months.

Bliss gained observational experience by setting up astronomical instruments on the old city walls near New College Lane, relying on a meridian mark on All Souls College for alignment. He started to send his observations of Jupiter's moons to Bradley and the two Oxford scholars continued to correspond for the next two decades. Bliss also secured valuable intellectual support and practical experience from George Parker, second Earl of Macclesfield, who invited the Oxford astronomer to help him observe a comet in February 1745. Parker had his own observatory at Shirburn Castle, about 20 miles (32km) south-east of Oxford, that was equipped by the same London craftsmen who supplied the Royal Observatory.

By 1761, Bradley was starting to suffer from poor health and asked his Oxford colleague to help with the observations at Greenwich. Bliss arrived at the Observatory in early summer and worked with Bradley's assistant, Charles Green, to observe the transit of Venus on 6 June with a reflecting telescope by James Short (24). After Bradley's death in July 1762, Bliss was appointed as the fourth Astronomer Royal and he continued to observe the Sun, Moon and planets with Green as his assistant. Sadly, his own health began to decline and he died suddenly on 2 September 1764. He was buried a few days later at St Margaret's, Lee – the same churchyard as Halley – just 1 mile (1.6km) away from the Observatory.

*Nathaniel Bliss*, British School, after 1736
(BHC4144)

# Nevil Maskelyne
## 1732–1811

On 25 July 1748, the teenage Nevil Maskelyne witnessed a total solar eclipse from his London home that inspired him to pursue a career in astronomy. Over the next 20 years, he nurtured both his academic and practical skills to fulfil his ambition. Firstly, he studied mathematics at Cambridge and became a curate to secure an income. Secondly, in 1758, he became a Fellow of the Royal Society, which brought him to the attention of the planning committee for the forthcoming transit of Venus expedition. The ambitious young astronomer was appointed and the three-month-long voyage to Saint Helena in 1761 provided him with an ideal opportunity to hone his practical skills in the lunar distance method (see *The quest for longitude*). Two years later, Maskelyne gained more navigational experience by travelling to Barbados for the second sea trial of Harrison's fourth timekeeper.

In October 1764, Maskelyne returned to London to discover that the Astronomer Royal, Nathaniel Bliss, had died and that he was the main candidate. He was eventually appointed on 8 February 1765 and set to work on devising the *Nautical Almanac*, focusing on critical observations of the Moon and certain bright stars. The survival of his observing suit is testimony to his active involvement with the 90,000 observations made over the 45-year course of his directorship. With responsibility for overseeing Board of Longitude publications, along with his Royal Society commitments in devising the scientific objectives of voyages of exploration, it was a demanding and exhausting role. Despite this, Maskelyne still found time to host sumptuous dinner parties that brought together a diverse mix of scholars and craftsmen. A fluent French speaker, he also networked widely with astronomers in Sicily, Germany, Poland, Russia and France, sometimes recommending London makers to supply new instruments.

Maskelyne died at Greenwich on 9 February 1811 after a short illness. In collaboration with others, he had fulfilled the Observatory's original remit in 'perfecting the art of navigation' and had firmly established Greenwich as an essential place of reference for both time and longitude, a legacy that continued to shape the Observatory's work for the next two centuries.

*Nevil Maskelyne*, John Russell, about 1802
(ZBA5100)

# JANUARY 1767. [9]

Distances of ☽'s Center from Stars, and from ☉ east of her.

| Days | Stars Names. | Noon. ° ′ ″ | 3 Hours. ° ′ ″ | 6 Hours. ° ′ ″ | 9 Hours. ° ′ ″ |
|---|---|---|---|---|---|
| 1 | | | | | |
| 2 | α Pegasi. | 46. 41. 15 | 44. 57. 51 | 43. 14. 53 | 41. 32. 32 |
| 3 | | 33. 15. 35 | 31. 40. 16 | 30. 6. 42 | 28. 35. 2 |
| 4 | α Arietis. | 57. 55. 16 | 56. 6. 21 | 54. 17. 44 | 52. 29. 25 |
| 5 | | 43. 32. 47 | 41. 46. 31 | 40. 0. 36 | 38. 15. 1 |
| 6 | | 62. 4. 49 | 60. 22. 21 | 58. 40. 17 | 56. 58. 37 |
| 7 | Aldeba- | 48. 36. 32 | 46. 57. 27 | 45. 18. 47 | 43. 40. 35 |
| 8 | ran. | 35. 37. 28 | 34. 2. 38 | 32. 28. 29 | 30. 55. 5 |
| 9 | | 23. 22. 20 | 21. 55. 18 | 20. 30. 0 | 19. 7. 3 |
| 10 | Pollux. | 51. 3. 14 | 49. 27. 59 | 47. 52. 57 | 46. 18. 9 |
| 11 | | 38. 27. 43 | 36. 54. 20 | 35. 21. 12 | 33. 48. 17 |
| 12 | | 62. 42. 22 | 61. 9. 30 | 59. 36. 47 | 58. 4. 13 |
| 13 | Regulus. | 50. 23. 35 | 48. 51. 53 | 47. 20. 18 | 45. 48. 52 |
| 14 | | 38. 13. 40 | 36. 43. 0 | 35. 12. 28 | 33. 42. 3 |
| 15 | | 26. 11. 51 | 24. 42. 9 | 23. 12. 34 | 21. 43. 10 |
| 16 | | 68. 17. 41 | 66. 48. 34 | 65. 19. 30 | 63. 50. 31 |
| 17 | | 56. 26. 28 | 54. 57. 51 | 53. 29. 15 | 52. 0. 41 |
| 18 | Spica ♍ | 44. 38. 16 | 43. 9. 50 | 41. 41. 25 | 40. 13. 0 |
| 19 | | 32. 50. 51 | 31. 22. 21 | 29. 53. 51 | 28. 25. 19 |
| 20 | | 21. 2. 16 | 19. 33. 33 | 18. 4. 47 | 16. 36. 0 |
| 21 | Antares. | 54. 40. 6 | 53. 9. 18 | 51. 38. 17 | 50. 7. 5 |
| 22 | | 42. 27. 36 | 40. 54. 57 | 39. 22. 2 | 37. 48. 50 |
| 20 | | 120. 36. 39 | 119. 14. 38 | 117. 52. 30 | 116. 30. 15 |
| 21 | | 109. 36. 50 | 108. 13. 39 | 106. 50. 14 | 105. 26. 38 |
| 22 | | 98. 25. 11 | 97. 0. 7 | 95. 34. 48 | 94. 9. 12 |
| 23 | The Sun. | 86. 56. 45 | 85. 29. 15 | 84. 1. 25 | 82. 33. 14 |
| 24 | | 75. 6. 56 | 73. 36. 29 | 72. 5. 38 | 70. 34. 23 |
| 25 | | 62. 51. 46 | 61. 17. 54 | 59. 43. 36 | 58. 8. 51 |
| 26 | | 50. 8. 25 | 48. 30. 56 | 46. 53. 0 | 45. 14. 36 |

## NAUTICAL ALMANAC (1767)

The newly appointed Astronomer Royal arrived at the Observatory in March 1765 with a clear mandate from the Board of Longitude (6): to produce an almanac to simplify and stimulate the use of the lunar distance method by mariners (see *The quest for longitude*). Maskelyne had already written *The British Mariner's Guide* (1763) explaining how to use Mayer's lunar tables (16) with a Hadley octant (27) for measuring angles between the Moon and stars to calculate one's position to within 0.5° longitude. However, the calculations were complex and took several hours to complete.

Within months, Maskelyne had devised the layout of the proposed *Nautical Almanac*. The core set of tables provided mariners with the predicted position of the Moon relative to certain bright stars, as seen from Greenwich, at 3-hour intervals for every day of the year. The book also featured information on calendars, measuring time by the Sun, the phases of the Moon, the orbits of Jupiter's moons, and the Sun's apparent position against the background stars (ecliptic). The accompanying *Tables Requisite* enabled mariners to correct their observations for the distorting effects of the Earth's motion (parallax) and atmosphere (refraction).

Working with just one assistant, Maskelyne used the Bradley transit telescope (17) to collect the necessary observational data which was then posted to a network of human 'computers' (22). They combined these figures with mathematical corrections to generate the predicted positions for the year ahead. The first edition of the *Nautical Almanac* was subsequently published on 6 January 1767. Longitude calculations at sea were now reduced to just 30 minutes' work and Maskelyne later extended the predictions so that long-distance mariners could buy almanacs for several years in advance.

The production of the *Nautical Almanac* became an essential function of the Observatory that cemented the institution's reputation as a key reference location. Similar almanacs were produced by observatories elsewhere but, by the 1850s, several countries had switched to using the *Nautical Almanac* for their own navies. Now published jointly by the UK Hydrographic Office and the US Naval Observatory, the renamed *Astronomical Almanac* is still carried by all Royal Navy ships and many other vessels today.

←
*Nautical Almanac* (first edition), Nevil Maskelyne, 1767 (ZBA5691)

→
Detail showing the bookplate with Maskelyne's signature (ZBA5691)

## 20
## MASKELYNE'S OBSERVING SUIT

Keeping warm is always a challenge for astronomers as they observe the stars on cold, clear nights. Maskelyne circumvented this discomfort by commissioning an observing suit made from silk, wool and linen. The trousers extend around the feet for maximum cover but we have no record of whether Maskelyne completed the outfit with gloves and a hat. Scuff marks, worn patches and tiny frayed edges indicate that he did indeed use it.

Intriguingly, the extra fabric around the middle and seat of the trousers clearly show that Maskelyne did not go hungry. The household accounts are filled with listings of regular deliveries of meat, dairy items and fresh produce from friends and family living on country estates. Similarly, the visitors' book reveals the regular occurrence of dinner parties at the Observatory, supplemented with luxury foods, wines, spirits, exotic teas, coffee and hot chocolate from various London suppliers.

For Maskelyne's assistants, life at the Observatory was less comfortable. In the 1790s, the young astronomer David Kinnebrook wrote to his father complaining about the wintry conditions at Greenwich, describing his struggle to work with frozen instruments in chilly temperatures. Battling against boredom and loneliness presented another challenge, as shown by this evocative description by Kinnebrook's successor, Thomas Evans: 'Nothing can exceed the tediousness and *ennui* [boredom] of the life the assistant leads in this place, excluded from all society, except, perhaps, that of a poor mouse.' Evans managed to escape this boredom in June 1798 by marrying and departing with Deborah Mascall, governess to the Astronomer Royal's daughter, Margaret (29).

Even with his luxurious suit, Maskelyne was acutely aware of the health risks of observing in the cold. His notebooks feature numerous recipes for cough and cold remedies made from kitchen ingredients such as lemon, ginger and honey, which will be familiar to many today.

Observing jacket and trousers, unknown maker, 1760s (ZBA4675, ZBA4676)

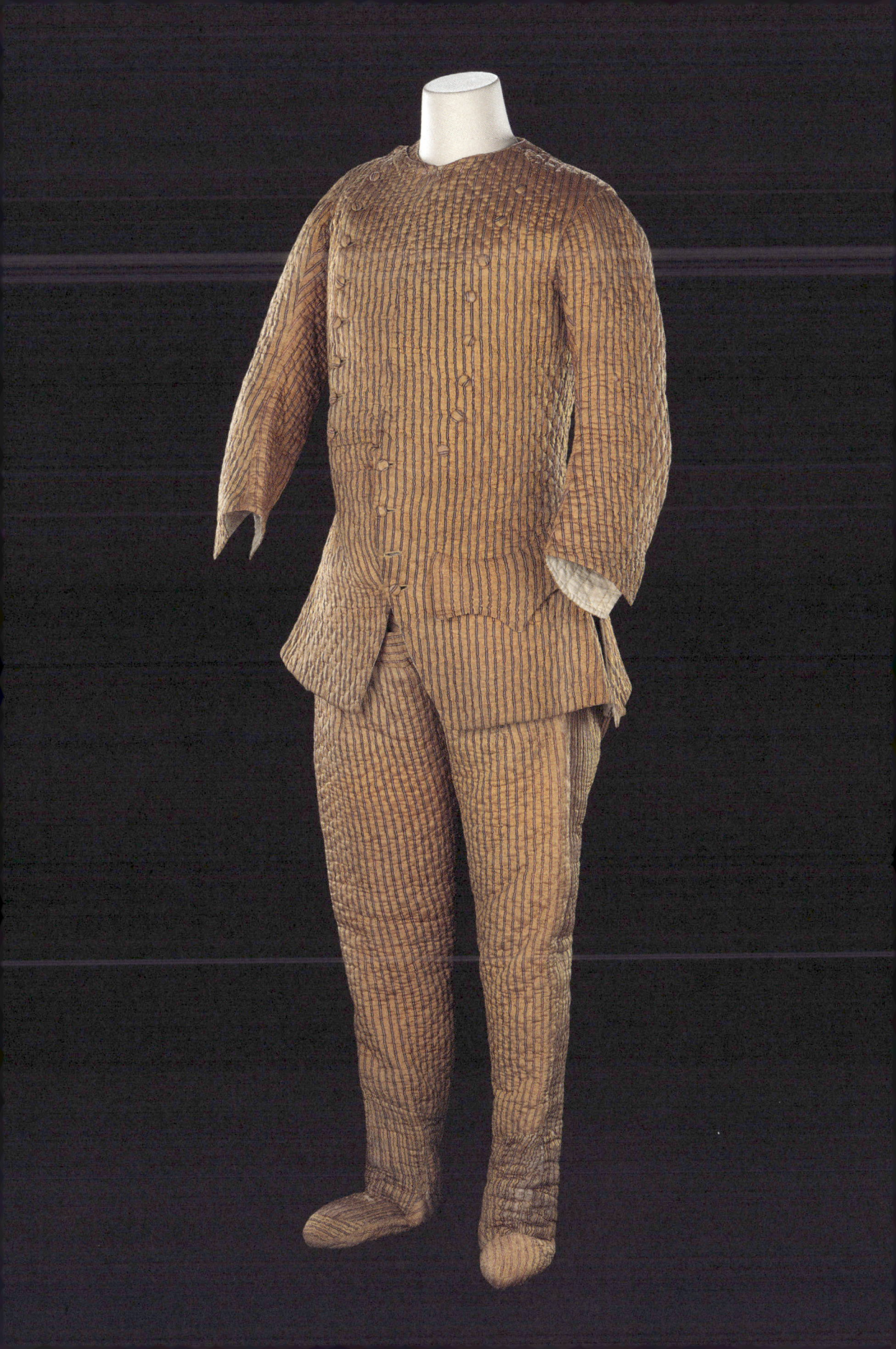

## 21
## MARINE TIMEKEEPER BY HARRISON, 'H4'

On 5 May 1766 Maskelyne collected Harrison's fourth marine timekeeper from the Admiralty and brought it to the Observatory, where it has largely remained ever since. Measuring just 5in. (13cm) in diameter and weighing only 3.2lb (1.45kg), the timekeeper represented a fundamental shift in Harrison's approach (13, 14, 18). In 1751 he had commissioned an innovative watch from John Jefferys that was more accurate than expected, leading him to seek development funding from the Board of Longitude. At the time, watches were notoriously inaccurate, often losing around a minute a day. By adding a heavy gold rim and creating a balance with a bigger swing that beat five times a second, Harrison produced a highly consistent watch, known today as 'H4'. He reduced frictional losses by adding bearings made from ruby and diamond and incorporated miniature versions of the bimetallic strip and remontoire from 'H3' (18) to improve the temperature compensation and driving force respectively.

The watch was eventually tested on voyages to Jamaica (1761) and Barbados (1764) and performed exceptionally well. The resultant error of just 39.2 seconds after 47 days at sea was well

Marine timekeeper 'H4', John Harrison, 1759, outer case with dial and movement (ZAA0037)

within the demands of the 1714 Longitude Act (6) but the Board refused to pay Harrison until he had disclosed his designs. He duly did so in August 1765, after which he received £10,000. In addition, the Board demanded that Harrison personally make two copies of 'H4', with a separate copy made by watchmaker Larcum Kendall ('K1'). All four original Harrison timekeepers were also required to undergo testing at the Observatory.

Having collected 'H4', Maskelyne ensured that the timekeeper was wound daily and checked against the Observatory's accurate clocks under the watchful eyes of both an Observatory assistant and an officer from Greenwich Hospital. But after ten months of variable results, the Astronomer Royal concluded: 'Mr Harrison's watch cannot be depended upon to keep the longitude within a degree in a West India voyage of six weeks.' It was a damning statement that stymied Harrison's efforts to secure his full reward but, after petitioning George III and Parliament, the clockmaker eventually received a final payment of £8,750, just three years before his death in March 1776.

## 22
## PAIR OF FLOOR GLOBES

Standing 3ft (1m) tall, these elegant 18in. (46cm) diameter globes were designed to showcase the latest discoveries in geography and astronomy. The terrestrial globe features the voyages of Captain Cook while the celestial version shows nearly 6,000 stars, clusters and nebulae (gas clouds), as observed by Maskelyne, Francis Woollaston and William Herschel (30).

In 1810, Maskelyne ordered smaller versions of the terrestrial globe to be sent to his four most trusted and longest-serving computers working on the Observatory's *Nautical Almanac* (19): J. Brown (Sheffield), Henry Andrews (Royston), Nicholas James (Saint Hilary) and Mary Edwards (Ludlow). Maskelyne envisaged that these globes would be used for lunar eclipse calculations and advised the recipients to annotate the globe with red ink that could be easily removed with a linen cloth.

Recruited via personal contacts of the Astronomer Royal, the computers were regularly sent parcels of mathematical books and observational results for determining astronomical data for the year ahead. Once their assigned months had been completed, each pair of computers sent their calculations to the comparer for checking. The value of Maskelyne's two-stage system became apparent in January 1770 when the Cornish comparer Malachy Hitchins deduced that two computers had copied each other's calculations, rather than working independently. The computers were re-employed several years later but they were never assigned the same month's calculations again.

One of Hitchins's most reliable computers was Mary Edwards, who officially began computing after her husband's death in 1784 but whose rapid uptake of the task seems to suggest her prior involvement. By 1789, Maskelyne was dependent on a highly efficient team of computers and comparers whose combined skills and accuracy enabled them to calculate predicted astronomical data up to a decade in advance. But the correspondence was not all work-related: Maskelyne also turned to Andrews and his wife for assistance in recruiting a nursery maid for his daughter Margaret (29).

Pair of terrestrial and celestial floor globes, William and Thomas Bardin, about 1800
(ZBA5106, ZBA5107)

Pl. 3.

Computations for finding the Longitude by Observations taken Sep. 6. 1767.

| Time by Watch | Dist. ☉ or ★ ☽ | Altitude ☉ or ★ | Altitude ☽ |
|---|---|---|---|
| H ′ ″ <br> 15. 0. 38 | ° ′ ″ <br> 44. 47. 51 <br> Pegasi | ° ′ ″ <br> 33. 9. | ° ′ ″ <br> 12. 59 |
| Mean | Mean | Mean | ☽ Mean ☾ |

| Computations continued. | | | |
|---|---|---|---|
| Horl. Parx. at Noon | 60. 11 | Propl. Logarithm | 4758 |
| At Midnight | 60. 34 | Do. | 4730 |
| Propl. Logarithm | 4758 | ½ Difference | 28 |
| Propl. part + or – | 26 | 11h Do. Propl. part | 26 |
| Propl. Logarithm | 4732 | = 60. 33. Horl. Parallax |  |
| Distance ☉ or ★ ☽ Observ'd | | | 44. 47. 51 |
| ☉ or – Semidiameters ★ ☽ | | | 16. 33 |
| Distance of the ☉ or ★ and ☽ centers | | | 44. 31. 18 |

For rectifying the Watch Observ'd at ....... 4. 59. 32
Altitude ☉ lower limb 10. 29. Watch by
Suppos'd Apparent time ........ 4. 59. 32
Latitude compd. 17. 30 S. Longitude compd. 64. 32.
From go. Co. Lat. 72. 30 + or – in time =p.6. 4. 18. 8
Time Observ'd by Watch ........ 4. 59. 32
Suppos'd Apparent time at Greenwich ........ 10. 24

Computation of Refraction by Mr. Lyon's Table.

Alt. ☉ corrd. 33. 6. 33. 13 T.N. 2021 Do. T.N. 2021
Alt. ☽. 13. 12. 34. 13. 2106 33. 14. 1916
T.N. 2021 1st Diffce. 85 2d Diffce. 205
+ 1st Propl. part 8 1st Prop. part 8 2d Propl. part 41
Sum 2029
– 2d Propl. part 41
to this 1988 Number + for an Index 2. 1988
+ Logarithm Co. Sect. Distance 44. 31. 10. 1542
Logarithm 226 — 2. 3530
By Tab. 2d with Dist. & less Alt. 115 Distance less 90. –more +
Sum or Difference is the 111 = 1· 51 Effect of Refrac".
Distance ☉ or ★ ☽ centers 44. 31. 18
Effect of Refraction 1. 51
Distance clear'd of Refraction 44. 33. 9.

For Computing the Apparent time.

☉ Declination Sep. 5th 6. 50. 10 N. 6. 50. 10
Do. 6 6. 27. 49
Difference in 24 hours 22. 21 = 1 —
☉ Declination at the time of Observation 6. 49. 10
+ or – from 90. is Polar Distance 96. 49. —
Altitude ☉ lower limb Observ'd 10. 29
+ Semidiameter less by dip & Refraction 8
Altitude ☉ center corrd. 10. 37
– From 90. is Zenith Distance 79. 23
Zenith Distance 79. 23.
Polar Distance 96. 49 Ar. Compl. Sine 0. 00308
Co. Latitude 72. 30 Ar. Compl. Sine 0. 02058
Sum 248. 42.
½ Sum 124. 21. = 55. 39 Sine 9. 91645
½ Sum – Zenith Distance = 44. 58 Sine 9. 84923
Sum 19. 78956
½ Sum is Co. Sine 38. 17 9. 89478
Doubled 2
Horary ∠ 76. 34 p.6 = 5. 6. 16
Time by Watch when the Altitude ☉ was taken 4. 59. 32
Difference is Watch slow by 6. 44

For Parallax.

Altitude ☉ or ★ corrd. 33. 6 ″
– Refraction p. 2. 1
Alt. ☉ or ★ corrd. 33. 5. Co. Sect. 10. 2629
Dist. ☉ or ★ ☽ clear'd of Refn. 44. 33. Sine 9. 8460
Propl. Log. Hor. Parallax 4732
Propl. Log. Arch 1st. 41. 07. 5821
Altitude ☽ corrd. 13. 12
– Refraction p. 2. 4
Alt. ☽ corrd. 13. 8. Co. Sect. 10. 6435
Dist. ☉ or ★ ☽ 44. 33. Tangt. 9. 9931
Propl. Log. Hor. Parallax 4732
Propl. Log. Arch 2d. 13. 59 1. 1098
Arch 1st. 41. 7
Princl. Effect of Parallax 33. 8 or Parallax in Distance
Distance cleared of Refraction 44. 33. 9
Princl. Effect of Parallax 33. 8
Distance clear'd of Principal Effect of Parallax 44. 00. 1
By Table 4th for second corrn. of Parallax. 21
Reduced Dist. clear'd of Refraction & Parallax 44. 0. 22

To Compute from the Observations above.

Time by the Watch when Dist. ☽ was taken 15. 0. 38
Watch being slow by 6. 44
Apparent time at taking the Distance ☽ 15. 7. 22
+ or – for Longitude from Greenwich compd. 4. 18. 8
Apparent time at Greenwich 10. 49. 14
Mean of the Observ'd Altitude ☉ or ★ 33. 9.
+ Semidiameter less by Dip
Altitude of the ☉ or ★ corrd. 33. 6:
Mean of Observ'd Altitude ☽ 12. 59
+ or – Semidiam. according which limb is Observ'd 13
Altitude of ☽ corrd. 13. 12

By the Ephemeris.

Dist. ☉ or ★ ☽ Noon at 9 45. 6. 40 Do. 45. 6. 40
Do. at Midnight 12. 43. 24. 24 Red. Dist. 44. 0. 22
in 3 hours 1st Diffce. 1. 49. 16 2d Diffce. 1. 6. 18
Proportional Logarithm of 1st. Difference 2455
Do. 2d. Difference 4338
Proportional Log. 1. 56. 49 Diff. 1883
+ hour of the 1st. Dist. 9. 4. 12
Gives Apparent time 10. 56. 40 at Greenwich
Apparent time 15. 7. 22 at taking the Dist. ☉ or ★ ☽
Difference 4. 10. 42 in time = p.6 62. 40. 30
is Longitude between the Place of Observation and Greenwich.

By the Ephemeris.

Semidiameter ☽ at Noon 16. 24 16. 24
Do. at Midnight 16. 30 ″
12 hours Difference 6 = 5
+ For Increase of Altitude ☽ p. 153. 4 16. 29
Apparent Semidiameter ☽ 16. 33
+ Semidiameter ☉
Sum of Apparent Semidiameter

N.B. Distance clear'd of Refraction { less 90. take the Diffce. of the two Arches } is Principal Effect of Parallax { Arch first greatest = contra + – from the Dist. clear'd of Refr. <br> More the Sum of the two Arches <br> By the Requisite Tables find the Parallax in Altd. ☽ & by Table 4th with Distance & Parallax in { Altd. Dist. 10 } Difference sec. corrn. Parallax.
which is to be + if distance is less 90. but more –
By the requisite Tables page 3. 4 & 5 with Appt. Altitude ☽ and Horl. Parallax gives the Parallax in Altde. 56.

By Robert Bishop.

Publish'd according to Act of Parlt. Decemr. 9th 1768.

## 23
## LONGITUDE CALCULATION SHEET

Having invested time and funds into the production of the *Nautical Almanac* (19), the Commissioners of the Board of Longitude were keen to see the book in widespread use. Within two months of printing in January 1767, the Board stipulated that copies of the *Nautical Almanac* should be made available across British, European and colonial ports. However, the uptake was seemingly slow and so by the autumn of 1768, both the Admiralty and the Directors of the East India Company made it a requirement for officers to be certified in the use of 'lunars' for measuring longitude, fuelling a growth in printed forms that guided mariners easily through the mathematics in a step-by-step process.

This example was created by Robert Bishop in 1768 but has been filled in with the values for 5 September 1767, possibly as a worked example. The mariner starts his calculations in the first column with local time measurements by the Sun and a watch, followed by the various angle measurements between the Sun, Moon and required stars using an octant or sextant (27). He then factors in corrections for the distorting effects of the Earth's atmosphere ('Computation of Refraction') and changing view of the Moon from different locations ('For Parallax'), before making the comparison with the predicted data ('By the Ephemeris') provided in the *Nautical Almanac* to determine the difference in time and longitude.

Bishop was a well-known author of navigational guides who described himself as 'many years a Master in the Royal Navy'. With extensive experience of sailing around Jamaica, Cuba and Hispaniola (Haiti), Bishop sold his books at the Jamaica Coffee House in London's Cornhill and worked with Maskelyne to develop an early version of this worksheet. He was one of several teachers specified by the Board of Longitude in 1768 for instruction in the use of both the *Nautical Almanac* and Hadley's octant and became the most prominent certifier for both Admiralty and East India Company officers.

Longitude calculation sheet, Robert Bishop, 1768 (G298:1/3)

IN FOCUS

# Transit of Venus expeditions

Although the astronomers at Greenwich were mainly focused on their local observations for timekeeping and navigation, some astronomical events such as eclipses and planetary transits were only visible from specific locations around the globe, requiring them to travel. These expeditions involved much planning, funding and preparation but it was worth the effort as astronomers could use these astronomical alignments to refine their values of planetary distances and orbits, which in turn assisted navigational calculations. More broadly, these rare but significant events also gave Observatory staff the opportunity to participate in groundbreaking science that created a long-lasting legacy of new departments and instruments at Greenwich. The expeditions to witness the transits of Venus in 1761, 1769 and 1874 all became pivotal moments in the Observatory's history.

**Transits of Venus**
Approximately once a century, occurring as a pair of events about eight years apart, the planet Venus appears to cross (transit) the Sun's disc. The first recorded observation of this orbital alignment was made by the Lancashire astronomer Jeremiah Horrox, who used Johannes Kepler's laws of planetary motion to predict its occurrence in 1639. Two decades later, the Scottish mathematician James Gregory proposed using the geometry of observing a planet from different locations on Earth to measure the distance to the Sun, but the idea gained little attention until it was developed further by Edmond Halley in the 1690s. Halley had already observed the transit of Mercury from Saint Helena in 1677 but he later realised that using the larger, closer planet of Venus would yield better results. With no opportunity remaining to see such a rare event during his lifetime, Halley urged his peers to keep sight of this challenge, stating in 1716, 'I recommend it to the curious strenuously to apply themselves to this observation.' He deliberately published his work in Latin to reach a wide audience. Over 40 years later, French astronomer Joseph-Nicolas Delisle responded to Halley's call to action by creating a map of the best observing locations and urging global collaboration. In Britain, the Royal Society and Halley's successors at Greenwich dutifully organised expeditions around the globe to observe subsequent transits in 1761, 1769, 1874 and 1882.

**How does it work?**
It takes Venus only 225 Earth days to complete its orbit around the Sun but 584 days for the planet to return to the same position between the Sun and Earth, a point known as inferior conjunction. In most conjunctions, Venus appears to pass slightly above or below the Sun. However, if the conjunction occurs near one of the two points where Venus's orbit crosses Earth's (in June or December), Venus may pass directly in front of the Sun. These rare alignments happen in pairs eight years apart, but each pair is separated from the next by more than a century. This orbital

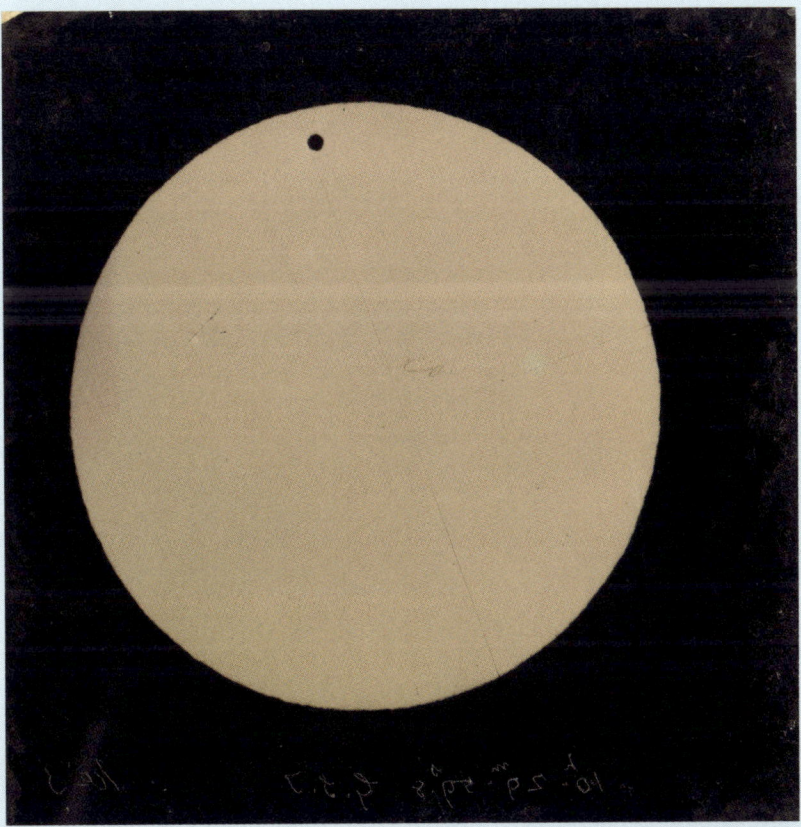

A square glass plate photopositive of the transit of Venus, unknown maker, 1874 (AST1083)

alignment gives us the opportunity to see Venus cross the solar disc over the course of six hours, either safely observed with professional filters or else projected onto a screen (never look directly at the Sun). The last pair of transits occurred in 2004 and 2012; the next pair will occur in 2117 and 2125.

**Why is it useful?**
If multiple observers from widely spaced locations on Earth can simultaneously observe the transit, we can use the geometry and time period of Venus' orbit to measure its parallax. This is the angular shift in the position of Venus seen against the Sun, which ultimately provides the distance between the Sun and Earth, known as the astronomical unit (au), and helps astronomers measure the scale of the Solar System.

The most important moments to measure are the four contact points when Venus begins to enter and leave the solar disc, although a visual phenomenon called the 'black drop effect', in which Venus appears to elongate into a teardrop shape, makes it very difficult to time these moments precisely.

**The transits of the 1700s**
The third, fourth and fifth Astronomers Royal all played a role in preparation for the pair of transits visible in the 1760s. Working with the Royal Society, the Astronomers Royal were responsible for choosing the observers and for procuring the necessary instruments from London's best makers, many of whom already supplied instruments to the Observatory. The Admiralty provided the transport and the astronomers relied on a network of observing locations chosen among colonial outposts, as recommended by the East India Company. The first event in 1761 was hampered by the Seven Years' War in which the British and French expedition teams were caught in the crossfire of battles at sea, but thankfully the hostilities ended in 1763 and the astronomers could once again appeal to national pride to secure funding for the second attempt in 1769. The Royal Society persuaded George III to fund two expeditions: one to Hudson Bay, Canada, and one to the Pacific island of Tahiti, which only became known to Europeans when HMS

*View from Point Venus, Island of Otaheite* [Tahiti], William Hodges, 1770s (BHC1938)

Transit of Venus hut erected on the Magnetic Ground, Edward Walter Maunder, 1874 (REG18/000454.34)

*Dolphin* visited in June 1767. Just over a year later, HMS *Endeavour* departed London for the eight-month voyage to Tahiti with Captain James Cook in command, accompanied by the former Greenwich astronomer Charles Green and the wealthy botanist Joseph Banks, who paid for his own passage as a private scientific expedition.

As a key member of the Royal Society's planning committee, the fifth Astronomer Royal, Nevil Maskelyne, devised a list of required instruments for each temporary transit observatory in both Tahiti and Hudson Bay:

- an astronomical quadrant for measuring the Sun and stars for local latitude
- reflecting telescopes (24), both for observing the transit itself and Jupiter's moons for longitude
- a micrometer attachment to create a series of vertical lines in the field of view for accurately measuring the diameter of Venus against the solar disc
- an astronomical regulator (accurate clock) for the timings.

After the event, Maskelyne was also responsible for collating the data from as many observers as possible, and for sharing the results with other astronomers via various scientific academies across Europe and the colonial states of America. With over 250 observers from 12 countries, it truly was a collaborative global effort, as advocated by Halley, and it contributed not only to astronomy but also to cartography, botany, zoology and geology, as the expeditions collected a wealth of scientific material.

### The transits of the 1800s

A century later, the seventh Astronomer Royal, George Biddell Airy, started to plan for the 1874 transit several years in advance. Firstly, he collaborated with telegraph companies to send time signals via submarine cables (61) to accurately measure the longitude of potential observing locations in Egypt. Secondly, he set up wooden observing huts within the grounds of the Observatory and installed a clockwork model on the roof of Flamsteed House to

help the observers practice their timings (65). By now, expedition travel was much easier and quicker by steamship and it took just two months for the British teams to reach Honolulu, Hawaii. The observing stations were well-equipped with the best instruments and technologies of the day:

- portable transit telescopes for measuring latitude via the Sun and stars
- 6in. (15.24cm) aperture refracting telescopes on equatorial mounts with clockwork drives to keep them in sync with the Sun's motion across the sky
- chronometers for measuring longitude
- new telescopes called 'photoheliographs' (64) that were specially designed for taking photographs of the solar disc

Astronomers were hopeful that photography could help define the key moments of contact but the observers struggled with the new technology and the resulting photographs were distorted, leading Airy to comment wistfully in his official report: 'I conceive it to be possible that some astronomer may yet think them worthy of rediscussion.'

### The results

Having analysed the data from the different observing locations in 1769, the Royal Society mathematician Thomas Hornsby announced a revised value of 93,726,900 miles (150,308,300km) for the Sun–Earth distance, which was a significant improvement on the highly variable results from 1761. For the next two centuries, astronomers continued to refine the value even further, both with the transit results from 1874 and 1882, along with additional measurements using the Moon, Mars and nearby asteroids such as Eros. In 1976, the International Astronomical Union agreed on a value of 149,597,870,700m (92,955,807.3 miles) for the astronomical unit, as measured by reflecting radar signals off the Sun and planets.

## GREGORIAN TELESCOPE

During the 1760s, Maskelyne and other astronomers were eagerly looking ahead to the 1769 transit of Venus (see *Transit of Venus expeditions*) to help measure the scale of the Solar System and to test Newton's theory of gravitation. Maskelyne was appointed to the Royal Society's planning committee in 1767 to review potential locations and to organise expeditions of observers and instruments.

This telescope, most likely made for a wealthy amateur, is similar to the one used by Captain Cook at the observing station in Tahiti. Maskelyne had recommended two reflecting telescopes by James Short but it seems that Cook also took his own telescope by Francis Watkins that he had previously used on his survey of Newfoundland in 1764. This type of telescope is known as 'Gregorian' after its inventor, Scottish mathematician James Gregory, who created his design in 1663. Light enters the optical tube at the open end and is reflected by a polished metal disc (speculum) onto a smaller mirror that directs the light into the eyepiece through a central hole. This configuration successfully reduced the distortion commonly seen with lens-based (refracting) telescopes, but Gregory's design was superseded by a similar version created by Isaac Newton that ultimately became the dominant design as it was cheaper and easier to make.

For the transit of Venus, the telescopes were mounted on top of a ballast-filled barrel for stability. Cook and his team also attached a set of precisely aligned wires (micrometer) by Dollond to help measure the diameter of Venus relative to the Sun. In his report to the Royal Society, Cook expressed his appreciation for the telescope, describing how the instrument's magnifying power (×140) enabled him to see the distinctive parts of the complete transit, despite the 'intolerable' temperature of 48°C (119°F) and the challenge of observing the notorious 'black drop effect', a phenomenon that astronomers continued to battle with a century later (64, 65).

◉ (and detail)
Portable Gregorian reflector telescope, Francis Watkins, about 1780 (AST0959)

## 25
## MASKELYNE'S COPLEY MEDAL

In 1775 Maskelyne was presented with the Royal Society's Copley Medal for 'his curious and laborious Observations on the Attraction of Mountains', making him the second of only three Astronomers Royal to receive this prestigious award. The obverse of the 1.7in. (43mm) diameter medal features the seated figure of Pallas surrounded by scientific instruments, while the reverse features the Society's coat of arms, along with its motto '*Nullius in Verba*' ('take nobody's word for it').

In his seminal book *Principia* (1687), Newton had proposed that a pendulum bob on a mountain would be deflected from the vertical by the gravitational attraction of the Earth's mass. Two French astronomers tried to measure this on the South American volcano Chimborazo in 1738 but the results were uncertain. Maskelyne took up the challenge in 1772 by appealing to the Royal Society for support. He was in luck: with spare funds and instruments from the transit of Venus expeditions, plus a leave of absence from the Observatory granted by the king, the project was approved. Maskelyne's assistant Charles Mason surveyed a range of possible peaks before choosing the conical, isolated Scottish mountain of Schiehallion, known locally as 'constant storm' on account of the continuous bad weather at its summit, over 2,533ft (1,083m) above sea level.

The Astronomer Royal arrived on 30 June 1774 and for the next four months was kept busy with numerous observations of select stars using a zenith sector (15). The instrument was vertically aligned with a plumb bob, meaning any perceived shift in the stars' positions, once other factors had been accounted for, would be a measure of the mountain's gravitational attraction on the bob. By observing the same stars on both the southern and northern slopes, and by assessing the fractional difference in latitude between the two locations by land survey, Maskelyne could calculate the mountain's density relative to the surrounding Earth. Four years later, mathematician Charles Hutton used these results to extrapolate a mean density of the Earth 4.5 times that of water, impressively close to today's value of 5.5.

Royal Society Copley Medal, John Sigismund Tanner, 1775 (ZBA2361)

## 26
## CHRONOMETER Nº.36 BY ARNOLD

The success of 'H4' (21) and its replica by Larcum Kendall showed that Harrison's design for a marine timekeeper was feasible, but it was still too complicated and expensive for widespread use. Other clockmakers took on the challenge of making marine timekeepers replicable and this example, 'Arnold 36', was a key milestone in the development of these navigational aids.

Originally from Cornwall, John Arnold began working on his ideas in his London workshop in 1767 after receiving a copy of *Principles of Mr. Harrison's Timekeeper* from Maskelyne. Arnold's first designs performed badly at sea but in 1775 he patented two innovations in the design of the balance (see *How does a mechanical clock work?*) that kept the timekeeper running uniformly across a range of temperatures. New models and patents followed in 1777. Arnold began producing watches with these features, and this instrument was the first in the new series, designated no.36 and finished with 18-carat gold.

Having informed the Board of Longitude of his improved design, Arnold submitted this timekeeper for trial at the Observatory for 13 months, starting in February 1779. The watch was measured when vertical, horizontal and face down and was even 'worn in the pocket' by the assistant who measured it daily, apart from two days in March when he forgot to wind it!

It performed exceptionally well, inspiring Arnold to publish its daily rate and error. At the same time, his ally Alexander Dalrymple, Hydrographer of the East India Company, published a pamphlet advocating the use of timekeepers at sea to improve navigation, boldly stating in a footnote that 'The Machine used for measuring Time at SEA is here named CHRONOMETER, my friend Mr [Joseph] Banks [President of the Royal Society] agreeing with me in thinking so valuable a machine deserves to be known by a name.' Arnold and other makers now began the race to make chronometers affordable and readily available.

Chronometer no.36, John Arnold, 1778
(ZBA1227)

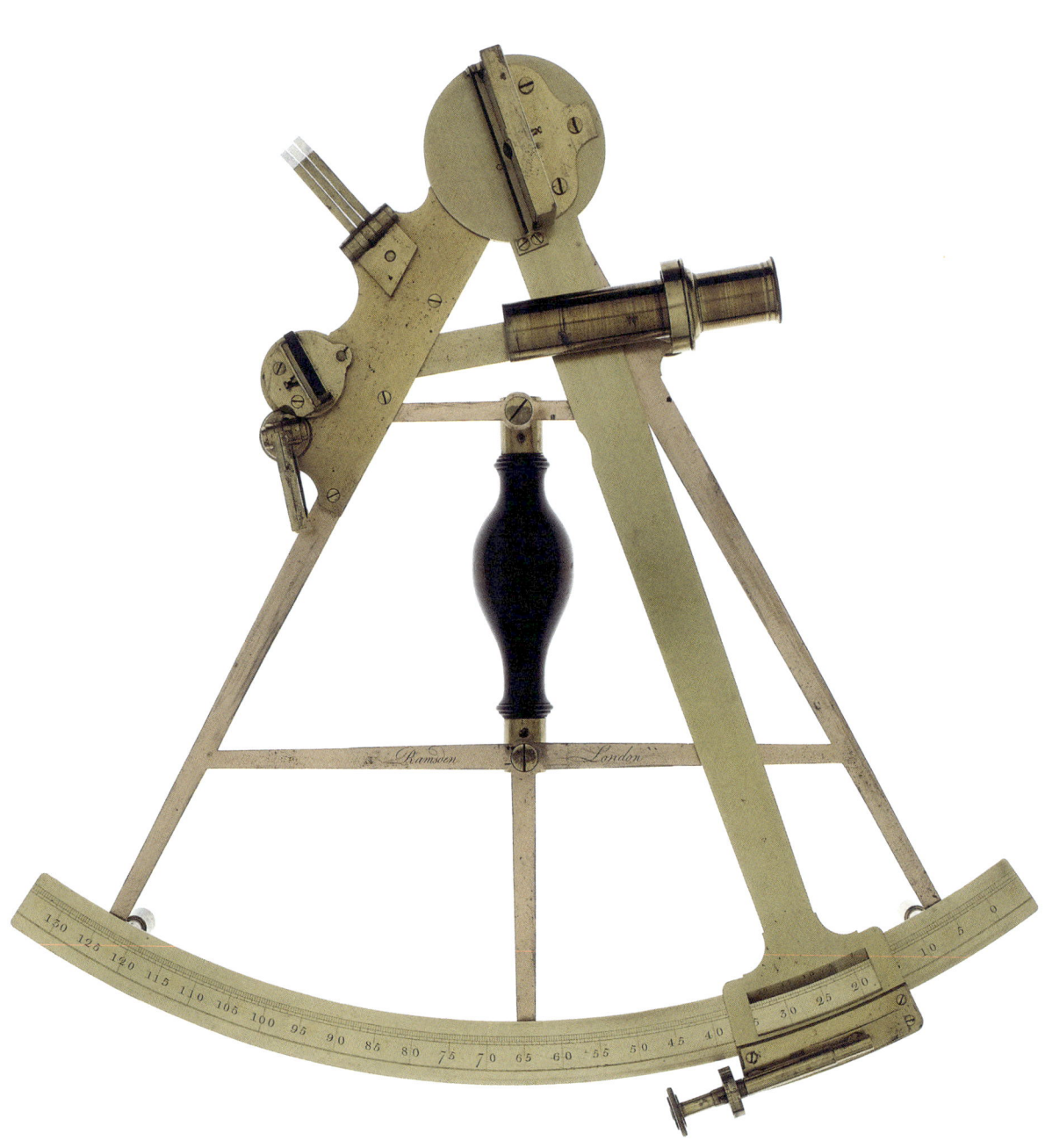

## 27
## SEXTANT BY RAMSDEN

By the 1780s, mariners had started to use sextants in conjunction with the *Nautical Almanac* (19) to measure longitude via 'lunars'. The instrument had emerged 50 years earlier when two makers created similar designs independently: Thomas Godfrey in Philadelphia and John Hadley in London, who presented his reflecting octant (one-eighth of a circle) to the Royal Society in May 1731. Hadley's instrument became the dominant design but despite encouraging trials at sea, the instrument's angular range proved to be too small for lunars (see *The quest for longitude*).

Further developments occurred in the 1750s when Tobias Mayer created a reflecting circle to be used with his improved lunar tables (16). Captain John Campbell tested the circle at sea and was inspired by the results to commission John Bird to transform Hadley's octant into a sextant (one-sixth of a circle) that could be used across the 120° range required for lunars. The first few sextants were large, heavy instruments that required a hip-held supporting pole for greater stability, but their navigational value became apparent after Captain Cook successfully used one on his first voyage to the south Pacific Ocean in 1768.

By the time Cook returned to London in 1771, the design and manufacture of sextants was changing, thanks to the innovative ideas of Jesse Ramsden. Based at 199 Piccadilly, Ramsden created a dividing engine that could engrave instrument scales mechanically, making the production of sextants quicker and more consistent. The Board of Longitude was highly impressed and paid Ramsden £615 to share the details with other makers.

The instrument shown here is typical of the 40 sextants made annually by Ramsden and his apprentices. Holding the instrument vertically, the navigator viewed the Sun or Moon through the telescope – relying on shades of red and green glass to protect their vision – and moved the metal arm along the scale until the mirrors brought the reflected view of the celestial body towards the horizon via double reflection. The instrument could also be used horizontally to measure the distance between two landmarks for surveying. In addition, this early version features a hinged cover used for checking the telescope's alignment, which was suggested by the Astronomer Royal himself and became known as the 'Maskelyne flap' in his honour.

Sextant, and detail of Maskelyne flap, Jesse Ramsden, 1783 (NAV1103)

Illustration from journal showing a sextant in use, John Lawrence King, 1846–47 (LOG/M/1)

IN FOCUS

# London's community of clock and scientific instrument makers, 1750–1800

During the late 1700s, the Observatory was a pivotal location in the exchange of knowledge between astronomers, clockmakers, scholars and instrument makers, many of whom promoted their association with Greenwich to secure lucrative contracts at other observatories across Europe. This collaborative relationship is most apparent in the visitors' book kept by Nevil Maskelyne, the fifth Astronomer Royal. By the 1780s, Maskelyne was regularly hosting dinner parties where guests could enjoy meat and game sent in from country estates and a well-stocked larder boosted by deliveries from local grocers. The visitors' book reveals a steady stream of scientific guests: astronomers, clockmakers, mathematicians, fellows of the Royal Society, Admiralty staff and merchants from the East India Company. At first glance, these dinner parties may seem trivial but, in reality, the good food and lively conversations were an important part of the symbiotic relationship between the Observatory and the city's thriving community of scientific instrument makers. Unsurprisingly, Maskelyne's diary also contains recipes for homemade remedies to relieve heartburn and indigestion.

**The demand for instruments**
With a growing empire and expanding trade routes, London's technical trades were thriving in the late 1700s with increased demand for instruments relating to navigation, weights and measures, surveying and ship construction.

When Maskelyne became Astronomer Royal in 1765, the capital was home to around 160 instrument makers, but this number had effectively doubled by the time of his death in 1811. Specialist makers established themselves in the relevant districts of the city: clockmakers in Clerkenwell, gunnery makers near the Tower of London and navigational instrument makers located alongside the river at Wapping.

In addition to these practical and economic demands, there was also a growing number of learned societies in London, composed of wealthy, well-educated individuals who wanted to have their own beautifully crafted philosophical instruments. Similarly, overseas traders and tourists passing through could admire and potentially purchase specialist instruments unavailable elsewhere. It was the epicentre of the instrument-making world and remained dominant for several decades.

**Training and recognition**
Becoming a scientific instrument maker in this period began at the age of 14 with a seven-year apprenticeship under the watchful eye of a Master of a Worshipful Livery Company. If successful, the young craftsman would progress to the next stage, becoming known as a 'Journeyman', before paying the fees and participating in the traditional 'taking the freedom' ceremony to qualify as a Master himself. Clockmakers were fortunate to have their own Worshipful Company established in 1631, whereas scientific instrument

Portrait of the optician John Dollond, Benjamin Wilson, about 1760 (ZBA0725)

'Little midshipman' figure, unknown maker, about 1780–1820 (AAB0173)

makers usually enrolled with the one most relevant to their expertise, such as the Goldsmiths for metalwork or the Spectacle Makers for optical instruments. It was not until 1956 that a dedicated Worshipful Company of Scientific Instrument Makers would emerge.

### Balancing work and family life

It is no surprise that, after living and working among their master's daughters for seven years, apprentices often married into the family. Like many other craftsmen and traders of the time, instrument makers and their families lived in a single space with little distinction between domestic and commercial areas. In addition to apprentices, makers would also enlist subcontractors to make specialist components. Sadly, the court records of the Old Bailey suggest that some of these temporary workers, now armed with a detailed knowledge of the premises, would later return as thieves and robbers. The more successful makers could afford a separate home and workshop, while others even invested in retail space on the Strand in central London to display their goods.

This interwoven relationship between work and home also created opportunities for women. So far, historians have traced over 58 women who were involved with the instrument trade in London during the period 1600–1800. Most worked behind the scenes by keeping the accounts and only became visible within the historical record after the death of their husbands. As widows, some women became more prominent as they continued the business by negotiating supplies, creating trade cards for advertising, and networking with other makers and organisations such as the Royal Society. The widows also continued to take on apprentices and paid the associated fees to the Worshipful Companies, even though they were usually not permitted to become members themselves.

# SOPHIA MASKELYNE'S DRESS

After 20 years at Greenwich as a bachelor, Maskelyne's life was about to change. Letitia and Sophia Rose, daughters of the deceased landowner John Pate Rose, lived a few miles away in Holborn. In July 1784, Maskelyne attended the marriage of his cousin, Reverend Sir George Booth, to Letitia and stood alongside Sophia as a witness. Within a month, Booth and his new wife were returning the favour as witnesses for the Astronomer Royal and Sophia Rose. The couple were married at St Andrew's, Holborn, on 21 August 1784, and this purple silk brocade dress is thought to have been Sophia's wedding dress. It would be another 50 years before Queen Victoria set the trend for white dresses. For Sophia, this was undoubtedly an expensive item and she made good use of it, as shown by the additional pieces of fabric to repair worn sections.

A year later, Maskelyne and his wife welcomed the arrival of their daughter Margaret (29), named after the Astronomer Royal's sister, Margaret Clive, widow of the controversial military commander and colonial governor 'Clive of India'. Yet while Maskelyne carefully recorded every stage of his daughter's development and education, his wife is only mentioned occasionally in his journals. Examples of her own writing are sparse, with one surviving letter to Caroline Herschel (30) offering us a glimpse of Sophia as the affable hostess: 'hoped you would have taken two dishes of coffee, and not gone till half-past eight'.

Our best source of information about Sophia Maskelyne comes from a series of portraits made at various points during the couple's married life, usually produced in pairs. The earliest portrait shows Sophia as a proud young wife with the infant Margaret on her lap and her wedding ring prominently on display. Later portraits demonstrate Sophia's love of fashionable clothes and jewellery; a miniature version even shows her wearing a matching miniature portrait of Maskelyne on a string of pearls. After her husband's death in 1811, Sophia moved with her daughter to the family home of Basset Down House, Wiltshire, where she lived until her death in 1821.

← Purple silk brocade dress, with a replica 'fichu' kerchief, unknown maker, about 1784 (ZBA4678)

→ *Mrs Sophia Maskelyne*, Mary Byrne, 1803 (ZBA5689)

## PORTRAIT OF MARGARET MASKELYNE

According to Maskelyne's journal entry, on 'June 25th 1785 at 20 minutes past one in the morning Mrs. Maskelyne was brought to bed of a girl, being her first child. July 26 she was christened Margaret the name of her god-mother my sister Lady Dowager Clive.' The Astronomer Royal was clearly delighted with the new arrival and, over the next few years, he studied her life in scientific detail, recording the dates of her weaning, first steps and inoculation against smallpox. He anxiously noted her symptoms during a bout of fever and wrote to colleagues for recommendations of a nursery maid.

Over a decade later, Maskelyne commissioned this fashionable portrait in which his daughter is shown with her beloved dog and their unique family home in the distance. Her teenage years were filled with a typical middle-class curriculum of literacy, numeracy, singing and dancing. Despite this hectic schedule, it seems that Margaret's governess, Deborah Mascall, still found time for romance: she later married one of Maskelyne's assistants, Thomas Evans (20), in June 1798.

Several surviving sketches by Margaret demonstrate her artistic talents. An incomplete view from the Observatory down to the Queen's House suggests that she may have used the newly invented camera lucida (lens and mirror device, 41) to project this panoramic scene onto paper. Other sketches and watercolours may have been produced freehand or else with the camera obscura (34) previously installed by her father within the north-west turret of Flamsteed House in 1778.

As Maskelyne grew old and became increasingly frail, Margaret became his secretary, using her skills in French and Italian to correspond with astronomers across Europe. She managed his affairs after his death and later moved to Wiltshire with her mother Sophia (28). In 1819 she married the barrister Anthony Mervyn Reeve Story, despite Sophia's concerns about the young man's finances. The couple's eldest son, Mervyn Story-Maskelyne was a pioneer in mineralogy and, in a subject that would have pleased his astronomical grandfather, became an expert on meteorites.

→
*Watercolour of the view from the Observatory,* Margaret Maskelyne, about 1810 (ZBA5098)

→→
*Margaret Maskelyne,* William Owen, about 1795–98 (ZBA5104)

# CAROLINE HERSCHEL'S DRESS AND BONNET

This dress and bonnet are believed to have been owned by Caroline Herschel, the first salaried female astronomer in Britain and discoverer of at least eight comets. Born in Hanover, she came to England in 1772 to join her brother William, who was working as a musician in Bath. This textile ensemble was acquired from the Herschel descendants in the 1970s and was reputed to have been worn by the famous female astronomer. Most notably, the dress's short length reminds us of Caroline's diminutive stature of just 4ft 3in. (129cm), believed to have been caused by childhood illness.

Alongside his lucrative music business, William gained a reputation as an accomplished telescope maker. As the house increasingly became a workshop, Caroline lamented how 'many a lace ruffle was torn or bespattered by molten pitch'. Life for the Herschel siblings changed dramatically upon William's discovery of the planet Uranus on 13 March 1781, inspiring George III to appoint him as 'the King's Astronomer' a year later. They moved to Slough to be closer to their royal patron and began their programme of 'sweeping' the skies to observe stellar gas clouds (nebulae) with increasingly powerful telescopes.

This proximity to London also increased the Herschels' visits to Greenwich, despite William describing Maskelyne as 'a devil of a fellow' after their first meeting. On one occasion, William demonstrated his planet-hunting 7ft (2.1m) telescope to Maskelyne, gleefully reporting afterwards that 'mine was found very superior to any of the Royal Observatory'.

For Caroline, William's £200 appointment included an additional £50 salary for her assistance in recording the observations. Over the next 20 years, she developed her own expertise in discovering new comets and always asked Maskelyne to verify her results, leading him to affectionately describe her as 'my worthy sister in astronomy'. This mutual respect was reinforced when, in 1798, Maskelyne encouraged Caroline to publish her list of 560 stars that had been omitted from Flamsteed's catalogue (7) and atlas (8). The following year, Caroline stayed at the Observatory for two weeks, later describing the Maskelynes as her 'esteemed friends'.

Silk dress with cotton bonnet, unknown maker, possibly 18th century (UNI3633, UNI3634)

## 31

## ASTRONOMICAL OBSERVATIONS (1798)

One of Maskelyne's most significant contributions to the work of the Observatory was to establish the convention of printing observational data on a regular basis to make it available to others. This protocol undoubtedly stemmed from Maskelyne's frustration in trying to secure the observations of one of his predecessors. When James Bradley died in 1762, his executors had simply removed 15 volumes of astronomical data, claiming that the work was his own property and would require compensation. This approach was not unique: both Flamsteed's beneficiaries and Halley's daughter had received payment for inherited astronomical manuscripts. The Board of Longitude tried to intervene but was advised to wait until Bradley's daughter, Susannah, turned 21. Sadly, she transferred the papers to her mother's family and the legal wrangles continued until 1776 when the manuscripts were presented to Oxford University for publication. The Savilian Professor of Astronomy, Thomas Hornsby, was assigned as editor but was hampered by poor health and so the first volume only appeared in 1798, followed by the second in 1805, which also contained the observations of Nathaniel Bliss, Bradley's brief successor.

The publication was certainly worth the wait: the volumes provided a wealth of detail on transit observations (17), the position of the Moon, zenith sector results (15) and a catalogue of 389 accurate star positions. Bradley was an accomplished observer who recognised the value of cross-checking his instruments for systematic errors, measuring the daily rates of the regulators (accurate clocks, 12) and taking barometer and thermometer readings to correct for atmospheric effects. These precautions ensured the ongoing use of Bradley's work as a highly valuable source of data. At first, it was used by Maskelyne to verify the latitudes and longitude difference between Greenwich and Paris observatories. It was also used in 1838 by Wilhelm Bessel at Königsberg Observatory to improve knowledge of star positions, a task that ultimately helped him measure the distance to the star 61 Cygni as 10.3 light years away, equivalent to 60 trillion miles (97 trillion km). This initial value was an important milestone in our bid to measure the scale of the Universe.

➔
*Astronomical Observations* (vol. 1), James Bradley and Thomas Hornsby (ed.), 1798 (PBG0607/1)

# ASTRONOMICAL OBSERVATIONS,

MADE AT THE

## ROYAL OBSERVATORY

AT

*GREENWICH,*

FROM

THE YEAR MDCCL. TO THE YEAR MDCCLXII.

BY THE

REV. JAMES BRADLEY, D.D.

ASTRONOMER ROYAL,

SAVILIAN PROFESSOR OF ASTRONOMY AT OXFORD,

FELLOW OF THE ROYAL SOCIETY,

AND MEMBER OF THE ACADEMIES OF SCIENCES AND BELLES LETTRES AT PARIS, BERLIN, PETERSBURG, AND BOLOGNA.

VOL. I.

OXFORD:
*AT THE CLARENDON PRESS,*
MDCCXCVIII.

## 32

## MODEL OF A CHRONOMETER ESCAPEMENT

This model of a chronometer escapement was the star witness in the battle for primacy between two of the most prestigious clockmakers in Georgian London. At the Board of Longitude meeting on 7 June 1804, Thomas Earnshaw and John Roger Arnold, representing his father John Arnold who had died in 1799, presented models and drawings of their respective spring detent escapements (see *How does a mechanical clock work?*). The designs replaced traditional pivots with a light spring, which reduced frictional losses and the need for oiling. Arnold's version was complicated and wore down quickly whereas Earnshaw's design was simpler and more robust. Earnshaw created his escapement in 1781 but could only afford to patent it in 1783, by which time Arnold had already patented his version a year earlier.

The 1780s had been a difficult decade for Earnshaw as he sought to relieve his financial debts by making chronometer production more profitable. He transformed the process by simplifying the design to just 128 parts and by dividing the labour among specialists around the country, ultimately making the instruments more affordable. By the 1790s, however, Arnold's patent had expired and Earnshaw's accusations of plagiarism grew louder, leading him to petition the Board for recognition. As a key Board member, the Astronomer Royal faced a difficult decision: both makers had worked on the Observatory's clocks, both had trialled their watches at Greenwich (26) and, in return, Maskelyne had recommended them to observatories elsewhere.

Matters came to a head in 1804 when the Board asked each maker to submit a model and detailed description to be circulated to other clockmakers for comment. Earnshaw's document was an outright attack on Arnold, prompting Sir Joseph Banks to publish a pamphlet in favour of Arnold followed by a counter-response from Maskelyne in support of the sharp-tongued clockmaker. Earnshaw's report was hastily destroyed and rewritten but, with 20 years having passed since the patent applications, coupled with Earnshaw's prickly reputation among his peers, community opinion remained divided. Consequently, the Board awarded £3,000 to each party. Despite this parity, it was Earnshaw's chronometer design that became the industry standard for the next century, as proudly featured in his portrait.

◉
Spring detent escapement model, Thomas Earnshaw, 1804 (ZAA0123)

◉
*Thomas Earnshaw*, Martin Archer Shee, about 1808 (BHC2674)

## 33
## PRINT OF THE EASTER FESTIVAL

Situated within Greenwich Park, the Observatory and its occupants witnessed many events associated with the town's cultural scene. This print from 1804 shows the main activities of the Easter Festival, a three-day event whose origins dated back to the twelfth century.

Young couples admire the view from the Observatory at the top of the hill, before preparing to roll down together (tumbling) with the enticing excuse to embrace in the chaos. Others at the foot of the hill use the opportunity to glimpse ladies' undergarments, as shown by the unfortunate lady revealing her red drawers. On the left, a crowd has gathered around two bare-knuckle fighting men, stripped to the waist. On the right, two young men distract a lady selling wares while another picks her pocket from behind the tree. In the foreground, an amorous sailor in red-striped trousers steals a kiss from a young woman in his clutches. For most, the fair was an opportunity for more innocent pleasures, such as strolling in the park and wearing fashionable bonnets for Easter.

Three decades after this print was made, the riotous behaviour seemingly continued, as described by Charles Dickens in his *Sketches by Boz* (1836). With his keen reporter's eye, Dickens brings the scene to life with the dancing, music, food stalls, fortune-tellers and performers, all of which left office clerks and apprentices to face the morning after 'with aching heads, empty pockets, damaged hats and a very imperfect recollection of how it was they did not get home'. While the numbers soared, particularly after the completion of the London and Greenwich railway in 1838, so did the trouble, leading to the Fair's closure in 1857.

*Greenwich Park, Easter Monday,* John Marshall and Co., 1804 (PAH3277)

102 A HISTORY IN OBJECTS

## 34

## PRINT OF THE VIEW FROM THE CAMERA OBSCURA

While the Easter Festival crowds (33) could only admire the city view from the Observatory gates, Maskelyne went further by installing a camera obscura in the north-western turret of Flamsteed House in 1778. Meaning 'dark room' in Latin, the concept of projecting light through an aperture in a dark space was first recorded in China over 2,500 years ago and later developed by Islamic and European scholars into the device used today. Flamsteed constructed a similar setup to safely observe the total solar eclipse of 2 July 1684 but the apparatus was later dismantled.

Maskelyne's device consisted of a mirror that reflected light onto a table below; a lens brought the view into focus. We have little insight into his motivation for reinstalling the device but it came into later use for his daughter Margaret (29) and for visiting artists such as Edward Pugh, whose landscape featured in Richard Phillips' guide *Modern London* (1804). In the foreground of the image overleaf we see the square tower of St Alfege's Church in Greenwich, while sailing ships throng the River Thames around Deptford. On the far right is a glimpse of one of the windmills along the western edge of the Isle of Dogs. Further away on the horizon are the twin towers of Westminster Abbey, the dome of St Paul's Cathedral and a myriad of church spires across the city.

Maskelyne's camera obscura was removed in 1840 by the seventh Astronomer Royal to make way for meteorological instruments. In 1994, a modern version consisting of an 8in. (200mm) lens and 9×12in. (228×304mm) octagonal mirror was installed in the Eastern Summerhouse, the same location as Flamsteed's original device over 300 years earlier. It shows a real-time image of the Queen's House projected onto a table 23ft (7m) below the mirror and remains a popular attraction for visitors today.

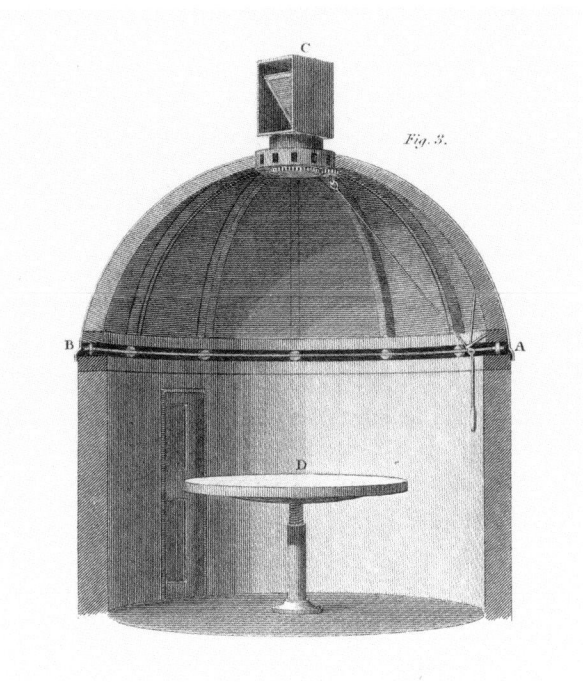

➲ 'Camera Obscura, Interior of a Typical Room', *A. Rees Cyclopaedia, Universal Dictionary of Arts and Sciences*, 1786 (PBD7574)

*The Cities of London and Westminster accurately copied from the table of the Camera Obscura in the Royal Observatory at Greenwich*, Edward Pugh, engraved by Tomlinson, 1 March 1809
(PAH2187)

JOHN POND
BORN MDCCLXVII
WAS ELECTED
ASTRONOMER ROYAL
MDCCCXI
WHICH OFFICE HE RESIGNED
MDCCCXXXV
AND DIED MDCCCXXXVI

# John Pond

## 1767–1836

With no known portrait and few surviving personal effects, John Pond, the sixth Astronomer Royal, is one of the Observatory's most elusive figures. Born into a wealthy London merchant family in 1767, he enrolled at Trinity College, Cambridge, in April 1784 where he explored his interests in chemistry, mathematics and astronomy but seemingly left without a degree. After travelling around the Mediterranean, he returned to England in the 1790s and settled in Westbury-sub-Mendip, where he constructed a private observatory.

Over the next decade Pond rose to prominence within the scientific community as a highly respected practical astronomer who formed a strong working relationship with the most prestigious instrument makers of the day. In 1807 he was elected Fellow of the Royal Society on the basis of his observational data that revealed discrepancies in the results provided by the ageing instruments at Greenwich. His work impressed the then Astronomer Royal, Nevil Maskelyne, who named Pond as successor just months before his own death in February 1811.

Once in post, Pond's reputation as an accomplished astronomer strengthened as he replaced the instruments with newer, better designs, along with innovative observational techniques that gave unprecedented results and culminated in a new catalogue of 1,123 stars. He also reinstated the previously neglected magnetic and meteorological observations and led the installation of the time ball in 1833.

But away from the telescope, Pond's tenure as Astronomer Royal was mired by bureaucracy and controversy. He successfully persuaded the Admiralty to fund an extra five assistants ('I want indefatigable hard-working and above all obedient drudges') to process the Observatory's data but his own workload increased significantly in 1821 when he was made responsible for the testing of Navy chronometers and, in 1829, for the management of the *Nautical Almanac*. With failing health, Pond faced growing criticism for his reputed lax management and negligence. In 1835 he became the first Astronomer Royal to resign from office and he died just a year later at his home in Blackheath near the Observatory.

John Pond's tomb, shared with Edmond Halley, at St Margaret's Church, Lee

## 35

## POND'S MURAL CIRCLE

In the 1780s, scientific instrument makers began to develop new techniques for creating circular scales that offered astronomers a bigger range and more accurate results than the quadrant-shaped instruments typically used at the time. Before he became Astronomer Royal, Pond voiced concerns about the accuracy of the instruments at Greenwich, leading Maskelyne to order this mural circle from Troughton in March 1807. Sadly, the Astronomer Royal died before the instrument was completed, leaving Pond as his successor to make the first observation with the instrument on 11 June 1812.

Troughton's circle consisted of a circular frame 6ft (1.8m) in diameter that was fixed onto a horizontal axis embedded within a granite pier aligned with the meridian. A telescope with a 4in. (10.2cm) aperture and a sequence of six micrometers – sets of vertical wires – spaced 60° apart on the pier, were used by the astronomer to measure the angle between the north celestial pole and certain bright stars as they crossed the meridian. Unlike quadrants which could only be used in one direction, the mural circle could be used twice as circumpolar stars crossed the meridian both north and south, about 12 hours apart. Troughton chose to make the circular scale from a tarnish-free, hard-wearing composite of palladium, gold and platinum, but he subcontracted the lenses to the opticians Dollonds on account of his colour-blindness.

Pond was initially pleased with the instrument's performance but within a few years he started to report errors caused by a slight movement of the horizontal axis within the pier. In 1824 he installed a second mural circle, designed by Thomas Jones, to provide simultaneous results for comparison. Jones' circle was intended for the Royal Observatory at the Cape of Good Hope but Pond's observations were so successful that the instrument remained at Greenwich until 1839 and another instrument was shipped to South Africa instead. Troughton's mural circle remained in use until 1850, when the building was reconfigured to accommodate the new Airy Transit Circle (47).

◉ (and detail)
6ft mural circle,
Edward Troughton,
Peter Dollond and John
Dollond, 1810 (AST0973)

## ASTRONOMER'S ALARM CLOCK

This unusual alarm clock was an innovative idea to help astronomers wake up at the correct time to view specific stars. Astronomers at Greenwich relied on measuring the exact moment at which certain 'clock stars' crossed the meridian to help monitor the accuracy of their mechanical clocks. These were bright, easily visible stars that had been observed by generations of astronomers, meaning their positions were precisely known as a reference.

At the start of each observing session, the astronomer inserted brass pegs into the drilled holes alongside the names of required stars. The clock gave a soft 'ping' every 10 minutes as the disc rotated and then produced a louder alarm to alert the astronomer when a pegged star was about to cross the meridian. This design was far more efficient than resetting a standard alarm clock after each transit. The sequence of star names around the dial matches the apparent motion of stars across the sky from east to west and includes the bright stars Aldebaran (in Taurus), Regulus (in Leo) and Spica (in Virgo). After each ring, the astronomer silenced the alarm by a lever and rewound the alarm train by pulling a cord.

The clock was designed by Observatory assistant Thomas Taylor, who was awarded a cash prize for his invention by the Society of Arts in 1819. Taylor was an important member of the Observatory staff, with responsibility for transit observations, data calculations and chronometer testing. Sadly, Taylor's behaviour deteriorated over the years, causing Pond to leave a cautionary description of his assistant in a statement written in 1835, claiming that 'his sight is imperfect; he has grown petulant and has latterly taken to drinking'. Upon his arrival, Pond's successor Airy tried to dismiss Taylor but, perhaps in recognition of his previous service in the Navy, the disgruntled assistant was pensioned off by the Admiralty instead.

Astronomer's alarm clock, Thomas Taylor and William Johnson, around 1815 (ZAA0525)

## 37
## RAIN GAUGE

Many observatories around the world today collect weather data as part of their daily work but this is a relatively recent phenomenon. The first few Astronomers Royal recorded limited temperature and atmospheric pressure readings using instruments positioned alongside their telescopes, but it seems to have been the fifth Astronomer Royal, Nevil Maskelyne, who adopted a more rigorous approach from 1 January 1807. Rainfall measurements did not commence until 1 December 1814, when his successor Pond began to use this grey painted brass can with a screw-in funnel made by Edward Troughton. The rain gauge was initially situated on a small plot of grass in the Meridian Courtyard but was moved to the Library roof in 1840. Water inside the cylinder was siphoned into a measuring vessel calibrated in inches. On average, the Observatory recorded about 25in. (63.5cm) of rainfall each year, with October and November the wettest months and February and March the driest.

Collecting meteorological data remained patchy at the Observatory until the creation of a dedicated Magnetic and Meteorological Department in 1840, headed by James Glaisher. Having endured months of rainfall while surveying the mountains of western Ireland, Glaisher was a passionate advocate for meteorology and campaigned to enhance the subject's reputation among the scientific elite. With his two assistants, he initially took hourly measurements of rainfall but this demanding schedule was soon replaced by self-registering instruments. Glaisher also used his contacts to recruit a national network of 60 observers who sent him local data for comparison. By the time he departed in 1874, the Observatory site had nine rain gauges and the data collection continued until its closure in the 1950s.

Rain gauge and funnel, Edward Troughton, around 1814 (NAV0815)

## CHRONOMETER N°.427 BY BROCKBANKS

Following Harrison's invention of the practical marine timekeeper, 'H4' (21) and its development by John Arnold (26), Thomas Earnshaw (32) and others, the marine chronometer was gradually adopted as an essential tool for navigation. The wealthy East India Company could afford to routinely purchase these expensive instruments but the cash-strapped Royal Navy could only prioritise chronometers for high-profile voyages. The situation changed when the Admiralty took over responsibility for funding the Observatory from the Board of Ordnance (1) in 1818 and made Pond 'Superintendent of Chronometers' in July 1821, tasking him with the testing, storing and distribution of the Navy's growing collection of about 130 chronometers. Pond was also required to instigate a series of annual trials in which clockmakers submitted their finest chronometers for testing against the Observatory's accurate clocks, in the hope of securing a commission from the Admiralty for the best-performing instruments. These annual 'Premium Trials' ran from 1822 until 1836 and fulfilled the Admiralty's objective of encouraging horological innovation to produce more accurate chronometers.

While Pond may have appreciated the extra £100 salary, testing chronometers was a demanding commitment that forced him to recruit extra staff and reallocate space within the Observatory, all of which detracted from his astronomical work. Yet not all chronometers were unwelcome guests, as shown by this example that remained at Greenwich for over 60 years. After a voyage on board the anti-smuggling patrol HMS *Egeria*, the timekeeper was repurposed for the Observatory's own use in January 1823. When the Admiralty requested its return six years later, Pond refused, claiming that it was 'advantageously employed in the Observatory; both in comparing the clock in the Chronometer Room with the Transit Clock, and for other purposes'. At some unknown date, the original octagonal mahogany case was modified to create a more convenient carry case. The instrument was mainly kept in reserve in the Computing Room but was occasionally used as a portable chronometer for astronomers wanting to check their observations of Jupiter's moons against the predicted timings of the *Nautical Almanac* (19). This useful adaptation meant the instrument remained in use until at least 1880.

◉ (and details) Marine chronometer no.427, fitted within a box with a carrying handle, Brockbanks, 1796 (ZBA1723)

## 39
## TIME BALL

Standing proud on the north-east turret of Flamsteed House, the time ball is one of the most recognisable features of the Observatory. This visual time-signal was the brainchild of Captain Robert Wauchope, who had created an experimental version at Portsmouth in 1829. He wrote to the Admiralty in June 1833 with his proposal to install a time ball at Greenwich so that mariners waiting on the Thames could check their chronometers before heading out to sea. The Admiralty and Astronomer Royal agreed and within a few months, mechanical engineers from the Lambeth firm of Maudslay, Sons and Field had installed the new device.

The original time ball consisted of a black leather canvas stretched across a 5ft (1.5m) diameter spherical wooden frame. One assistant cranked up the ball while another checked the time against regulator no.3 by Graham and triggered the release when required. Pleased by the success of the Greenwich system, Wauchope promoted his idea to the East India Company who installed time balls in ports across Mauritius, South Africa and India. By the 1850s, the time ball at Greenwich was controlled automatically by electric impulses from the Observatory's Shepherd Motor Clock (49), the telegraphic connections of which were also used to control other time balls across London and Kent (55). In 1919, the time ball was completely overhauled with a new winch system and aluminium sphere provided by E. Dent and Company, which remains in use today.

Apart from some interruptions caused by storms, technical failure and wartime circumstances, the time ball has dropped daily since 1833 with the same sequence: it rises halfway up the mast at 12.55 p.m., before progressing to the top at 12.58 p.m. It drops at precisely 1 p.m., coming to rest gently thanks to a cistern of compressed air. The astronomers chose 1 p.m. as the time of release because midday was an essential moment in their observations and calculation of Greenwich Mean Time (GMT). Now controlled by signals from Global Positioning System (GPS) satellites, the time ball continues to operate today and is released at 1 p.m. all year round. If you look closely, you will notice the dimples caused by workmen rolling the ball across the courtyard during restoration work in 1959.

The time ball at rest,
E. Dent and Company,
1919 (ZBA2245)

# Sir George Biddell Airy

## 1801–92

As the sixth Astronomer Royal, John Pond, suffered bouts of poor health and faced criticism for alleged inaccuracies within the *Nautical Almanac*, members of the Observatory's Board of Visitors – a group of expert advisors from academia and learned societies – began to search for a successor. With an established reputation as Director of Cambridge Observatory, George Biddell Airy was an obvious choice. He started work at Greenwich on 1 October 1835 and was joined by his wife Richarda and their three young children a few months later, once additional rooms had been added to Flamsteed House. Unlike Pond, Airy was not an observational astronomer but preferred to focus on the administrative and technical objectives of the Observatory's work. He was driven by a strong sense of national duty in providing the best astronomical data for practical purposes and successfully petitioned the Admiralty to invest in new instruments, most notably the Altazimuth (1847), the Merz refractor (1859) and the Airy Transit Circle (1850), many of which he designed himself. In contrast to some of his predecessors, he also ensured that all observations were published, along with an annual report for accountability.

During an intense period of scientific change, Airy kept pace by ensuring that the Observatory adopted new subjects and technologies such as telegraphy, solar photography, spectroscopy, meteorology and studying the Earth's magnetic field. As a government scientist, he served on various commissions and organisations and explored his broader interests in optics, engineering, measuring the density of the Earth, and improving the design of pendulum clocks.

Beyond his technical expertise, Airy applied his methodical mind to standardising the Observatory's work through new forms and procedures to ensure that his staff were disciplined and consistent in their tasks. The sheer volume of his surviving papers – extending over 360ft (110m) in the archives – also confirms his reputation as a workaholic. But behind closed doors in Flamsteed House, he was a devoted family man who worked alongside them in the evenings for 'taking the edge off his work'. He retired in August 1881 and moved to a house on the edge of Greenwich Park where he died after surgery in January 1892.

←
*George Biddell Airy*, M. & N. Hanhart, Maguire, Thomas Herbert, Ransome, George, 1852 (PAG6616)

*The Observatory, Cambridge*, Richard Bankes Harraden and E.F. McCabe, 1825
(PAH6090)

# ENGRAVING OF CAMBRIDGE OBSERVATORY

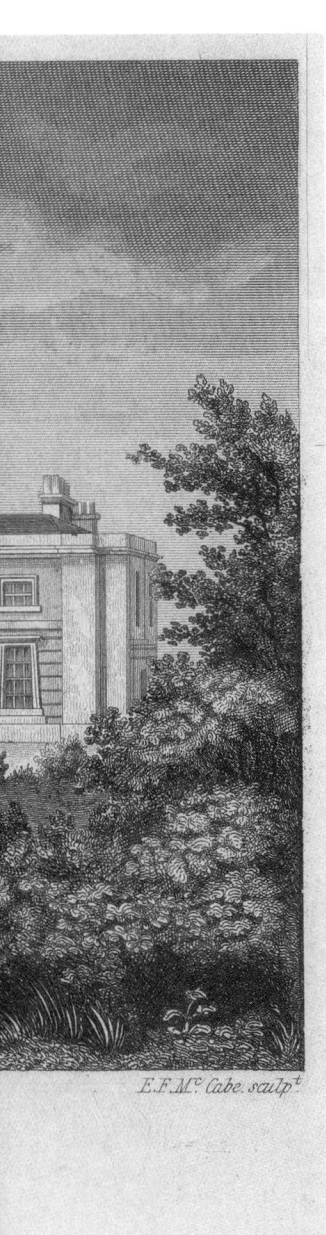

Airy's appointment as the seventh Astronomer Royal stemmed from his previous experience as Director of Cambridge Observatory, which was a new facility that had arisen in the 1820s as reformers tried to persuade their traditional-thinking peers to embrace exciting new European approaches in mathematics. A decade earlier, a progressive group of undergraduates had created 'The Analytical Society' to promote the distribution of translated French mathematics books. These radicals argued that astronomy could be used to showcase new mathematical techniques and their efforts were finally rewarded when the university agreed to construct an observatory in 1822. A year later, the neo-classical style observatory came into operation under the directorship of Robert Woodhouse, who continued until his death in December 1827.

With outstanding academic honours and formal recognition as Lucasian Professor of Mathematics, Airy was the obvious choice as successor. He took up residence in April 1828 and embarked on a systematic programme of observations using the 10ft (3m) Troughton and Dollond transit telescope, alongside a busy lecture schedule on topics such as optics, mechanics and gravitation. Five years later, Airy embarked on the design of a 12in. (30cm) aperture telescope – named after the Observatory's benefactor the Duke of Northumberland – that was installed on an English-style equatorial mount in a purpose-built wood and zinc dome. With his trademark attention to detail, Airy carefully designed a winch system to enable the astronomer to move the open dome without leaving his observing chair.

As more funding became available, Airy recruited three assistants, two of whom would later join him at Greenwich. He also instigated several practices that he continued as Astronomer Royal, namely keeping a daily 'Observatory Journal', printing template forms to reduce errors in the data reduction and publishing an annual report for distribution to other observatories.

But Airy's motivation to succeed at Cambridge Observatory was not based purely on professional ambition. With a guaranteed annual salary of £500, he was now financially secure and able to marry. In November 1829 he travelled back to Edensor, Derbyshire, where he had met Miss Richarda Smith five years earlier. She accepted his proposal and the couple were married in March 1830.

Playford — From a sketch by M.rs Airy

## 41
## SKETCH OF PLAYFORD COTTAGE

Once appointed as Astronomer Royal in June 1835, Airy divided his time between Cambridge and Greenwich, with Richarda and their three young children (George, Arthur and Elizabeth) eventually joining him to settle at Flamsteed House by the end of the year. The Airys enjoyed their new home but they also appreciated escaping from the whirlwind of city life to holiday in the rural village of Playford, near Ipswich, for several weeks each winter and summer. It was a place of many happy memories for Airy as he had frequently stayed there on his uncle's farm during his teenage years. With a well-stocked library, Arthur Biddell encouraged his nephew to read widely across classical history, poetry, mathematics and mechanics, inspiring the young scholar to continue his studies at Cambridge University. Biddell also introduced Airy to local contacts who would shape Airy's interests in later life, from the engineering family of Robert Ransome and sons, who would construct several large telescopes for Greenwich (47, 76), to the slave-trade abolitionist Thomas Clarkson MP, who lived nearby.

As an adult, Airy cemented his association with Playford by purchasing 'Church Cottage' in 1845. The Suffolk retreat became an essential part of family life and was captured on paper by Mrs Airy around 1850. Richarda's 'skill and fidelity in sketching', so admired by Airy, is evident in the intricate detail and nuanced shading. He himself enjoyed sketching with the aid of a camera lucida, a portable gadget in which the image of an object seen straight ahead is projected onto the artist's paper below, making it easy to trace. Both Airy and his wife would eventually be buried at Playford Church, just behind the tree in the sketch, alongside their teenage daughter Elizabeth and young sons George and Arthur who died tragically from childhood illness. Somewhat fittingly, the cottage passed down the generations to the granddaughter Anna Airy who continued Richarda's artistic legacy with a career as a pioneering female war artist in 1918.

◄
*Playford – From a sketch by Mrs Airy*, Richarda Airy, about 1850
(PAH6035)

►
Instruction leaflet for the camera lucida, unknown maker, about 1807 (NAV0516.2)

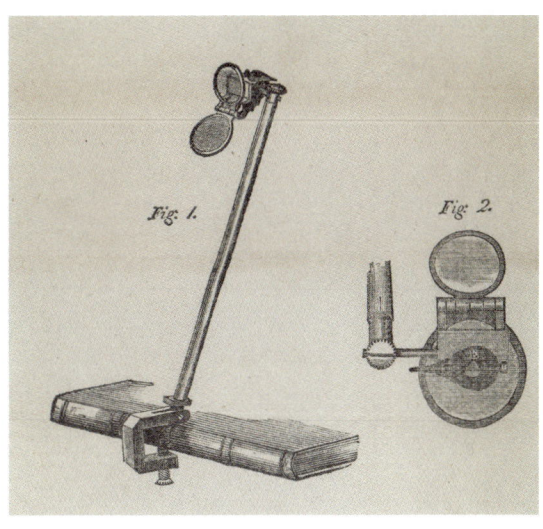

## 42
## AIRY'S HOLE PUNCH

After inheriting some of Pond's assistants, Airy set about creating his own hierarchy of observers (Assistants) and office staff (Computers). By creating a 'skeleton' form that transformed the complex mathematics of data analysis into a step-by-step process of basic arithmetic, Airy could fill his Computer roles with teenage boys on low pay while the more expensive Assistants focused on observations. Although some recruits advanced into the Assistant grade, most left for jobs elsewhere after their gruelling experience of long hours and pitiful wages.

For others, such as Edwin Dunkin, the Computer role was a useful entry point to a lifelong career in astronomy. Dunkin arrived in 1838 and retired 46 years later as Chief Assistant. In 1862 he described the business-like atmosphere of the Computing Room: 'Let us look over the shoulders of some of the assistants, one computer, generally a junior, is entering the observations made on the preceding night, another is employed on an advanced stage of the computations[…], while a third is arranging in order for binding, some of the manuscripts and letters of the preceding year.'

There was certainly plenty of paper to be bound. Around 1837 Airy began organising papers with a simple four-hole hand punch. As the number of Computers increased, so did the volume of papers. By 1841, Airy's staff was number-crunching its way through 18,000 skeleton forms of lunar data every year, prompting him to commission a six-hole punch from the engineering firm of Ransomes and May in 1843. Impressed by the result, Airy arranged for additional punches to be sent to the observatories of Cape Town, Melbourne and Pulkovo. This effective instrument for punching foolscap-sized papers (13×8in. or 33×20cm) became an essential part of the Observatory's daily routine. Airy's design was a precursor to the more familiar office gadget that emerged in the 1880s, making this one of the earliest known hole punches in the world. Its survival is a fitting reminder of Airy's zeal for keeping records and organising papers, as revealed by his 1860 instruction: 'No paper whatsoever to be destroyed.'

➔
6-hole paper punch,
Ransomes and May,
1843 (AST1169)

# 43
# GRAND ORRERY

This mechanical model of the Solar System is a superb example of the eighteenth-century fashion for displaying one's wealth and knowledge of astronomy. The first geared 'planetarium' of this type was made by clockmakers George Graham and Thomas Tompion in 1704. These popular teaching devices became known as 'orreries' after Charles Boyle, fourth Earl of Orrery, who commissioned and popularised some of the earliest examples.

The outer sides of the polygonal drum feature oval plaques showing the signs of the zodiac, while the interior contains a complex set of gears that powers the planets in their steady progress around the Sun on four moveable wooden circles, painted sky blue with gold stars. The tiny ivory globes can be set to move at their relative speeds, with Mercury circling the Sun 49 times for every single orbit of Jupiter, mimicking their orbital periods of 88 days and 11 years. Saturn progresses slowly as the outermost planet, prior to the discovery of the seventh planet Uranus in 1781 by William Herschel in Bath.

By Airy's time, planet hunting had become a contentious topic. In 1843, Cambridge scholar John Couch Adams began to calculate the position of an eighth planet thought to be disturbing the orbit of Uranus. He sought advice from Airy but a combination of missed opportunities and other competing demands delayed progress. On 23 September 1846, Johann Gottfried Galle at Berlin Observatory announced the discovery of the eighth planet, later known as Neptune, at the position predicted by French mathematician Urbain le Verrier. In a period of intense rivalry between British and European scientists, Airy faced criticism for the lost opportunity. But for the Astronomer Royal, the case was clear: the Observatory's remit was to provide data for timekeeping and navigation, not speculative planet hunting; he was merely helping the young scholar out of interest. Adams was eventually recognised alongside Le Verrier but the controversy haunted Airy's reputation for decades.

➔
Grand orrery with clock mechanism, unknown maker, 1750–1800, with clock mechanism, James Simonds, about 1840
(AST1067)

44

## REGULATOR N⁰.587 BY DENT

Standing majestically at 6.5ft (2m) tall, this carved mahogany case contains an astronomical regulator that truly epitomises the close relationship between astronomers and clockmakers. The timekeeper was made in 1842 by Edward J. Dent who was already well known at Greenwich for supplying Admiralty clocks and for submitting marine chronometers to the annual trials (38, 51), one of which was awarded first prize in 1829. The sixth Astronomer Royal, John Pond, described Dent as 'among the best workmen of the present day' and his successor seems to have agreed with him.

Around 1840, Dent was commissioned construct a suite of clocks for the newly built Pulkovo Observatory, situated 12.4 miles (20km) south of Saint Petersburg. To account for the extreme variations of the Russian climate, the clockmaker installed a mercury-filled pendulum that only varied in length by a small amount across changing temperatures. He also added a special type of escapement (see *How does a mechanical clock work?*), designed by Airy himself, to ensure the clock maintained a steady rate. Dent wrote to the observatory's director, Otto Struve, in October 1842 to reassure him that the clocks were due to be shipped imminently but, for some unknown reason, this regulator was not included in the final consignment. It remained in the craftsman's workshop and was later displayed alongside Dent's other innovative designs at the Great Exhibition in 1851.

Despite the unfulfilled order, Airy and Dent continued to work on other projects, most notably the design of the public clock for the new Houses of Parliament, officially called the Westminster Clock but commonly known today as 'Big Ben'. Sadly, Dent died before installation in 1859 but Airy continued to oversee the clock's function by installing a telegraph line between Westminster and Greenwich to check that the nation's most famous clock kept accurate GMT to within a second. Airy continued to pursue his horological interests with Dent's stepsons Frederick and Robert, culminating in the 1870s with the design of a regulator whose pendulum was automatically adjusted for changes in both temperature and barometric pressure through a combination of mercury and magnets. While Airy may have been Astronomer Royal, it was in the detailed design of clocks and telescopes that he excelled and where his interests lay, as opposed to making observations himself.

◥
Astronomical regulator no.587, Edward J. Dent, about 1842 (ZAA0605)

## 45

### DAYLIGHT VIEW OF THE GREAT COMET OF 1843

This artwork encapsulates some of the intellectual and social connections between the Royal Observatory and its sister organisations in South Africa and Scotland. Painted by astronomer Charles Piazzi Smyth while working at the Royal Observatory, Cape of Good Hope, this vista of Table Mountain shows the dazzling magnitude and extent of the Great Comet of 1843. As the sun-grazing comet made its closest approach to the Sun on 27 February, observers in the southern hemisphere were afforded the earliest and brightest views.

At that time, Smyth was working alone at the Cape Observatory, continuing with the routine meridian observations while the Director, Thomas Maclear, and his assistant, William Mann, undertook a survey of the Cederberg Mountains. Smyth recorded his comet observations and sketches in a journal that would later form the basis of this painting. The viewer is looking west, a few hours before sunset, to see the comet's central nucleus and lengthy tail of gas and dust, still clearly visible under the Sun's daytime glare. Over the next few weeks, Smyth meticulously recorded its lengthening tail and dimming magnitude, despite the meagre pair of telescopes available and many evenings lost to bad weather.

Smyth returned to Britain in 1846 and took up office as Astronomer Royal for Scotland at the observatory on Edinburgh's Calton Hill, devoting his energies to establishing the time ball (1852) and 1 p.m. gun signal (1861). He also investigated his wide-ranging interests in Egyptology, spectroscopy, cloud formations and testing mountain top locations as prospective observatory sites. Despite their common roles, Airy was quite dismissive of Smyth's ambitious and unconventional ideas. However, he was still prepared to act as peacemaker after a dispute about double stars between Smyth and fellow Scotsman J.D. Forbes at a British Association for the Advancement of Science (BAAS) meeting in Edinburgh in 1850.

More generally, the relationship between the observatories at Greenwich and the Cape remained strong for another century with a steady exchange of people, instruments and data. In particular, the continuation of solar and photographic observations at the Cape and observatories elsewhere during the Second World War helped Greenwich enhance its own patchy and disrupted set of data.

◄
*Daylight View over Table Bay Showing the Great Comet of 1843*, Charles Piazzi Smyth, 1843
(BHC4147)

IN FOCUS

# The Age of Magnetism

First used by Chinese mariners in the medieval period, the magnetic compass was a fundamental part of the navigator's toolkit by the time the Observatory was founded in 1675. Magnetism barely featured within the institution's original workload but this began to change during the early 1800s. In the wake of the Napoleonic Wars, the Royal Navy sent out its superfluous men and ships to survey the globe on behalf of its Hydrographic Office. One particular region of interest was northern Canada and the search for the North-West Passage, a potential network through the arctic icefields that could open up new trade routes with Asia. Other voyages headed to the extremes of our planet in a bid to locate the north and south magnetic poles. It was an exciting time for new expeditions and evolving theories about magnetism, as European explorers and scholars tried to survey and understand this fundamental force of nature. This intense period of interest – sometimes known as the 'Magnetic Crusade' or 'Magnetic Fever' – was responsible for a new observation programme at Greenwich that would endure for another 80 years.

**Early magnetic studies at Greenwich**

Understanding the nature of the Earth's magnetism was a challenge that had already occupied the second Astronomer Royal, Edmond Halley, long before he came to Greenwich. In the late 1690s he commanded two Royal Navy voyages to measure magnetic declination around the globe, tasked with determining how much a compass needle varies from pointing true north, as compared against the north star Polaris. His subsequent sea chart showing lines of equal declination (isogones) was the first of its type and Halley hoped that mariners could identify their location by matching the variation on their compasses to the relevant zone on the chart. But with large areas of equal variation, the results were too vague for accurate navigation, especially as the lines drifted in response to fluctuations in the Earth's magnetic field and the charts became outdated.

Magnetic observations at Greenwich remained intermittent until the creation of the systematic observation programme by the sixth Astronomer Royal, John Pond. In March 1816, the Admiralty requested that the Observatory start collecting magnetic data, possibly in support of the Navy's impending voyages to find the North-West Passage for 'the advancement of geography, navigation and commerce'. In response, Pond established a wooden magnet house in a hollow in the Observatory's lower garden where, from June 1818, he started to take daily observations of declination every four hours between 8 a.m. and 8 p.m. The observations ceased at the end of 1820, possibly as resources were redirected to the new programme of testing of chronometers (38), and by 1824 the magnetic building was reportedly unstable due to a landslip.

**Matching scientific and national interests**

For the next decade, magnetism was once more absent from the Observatory's remit but

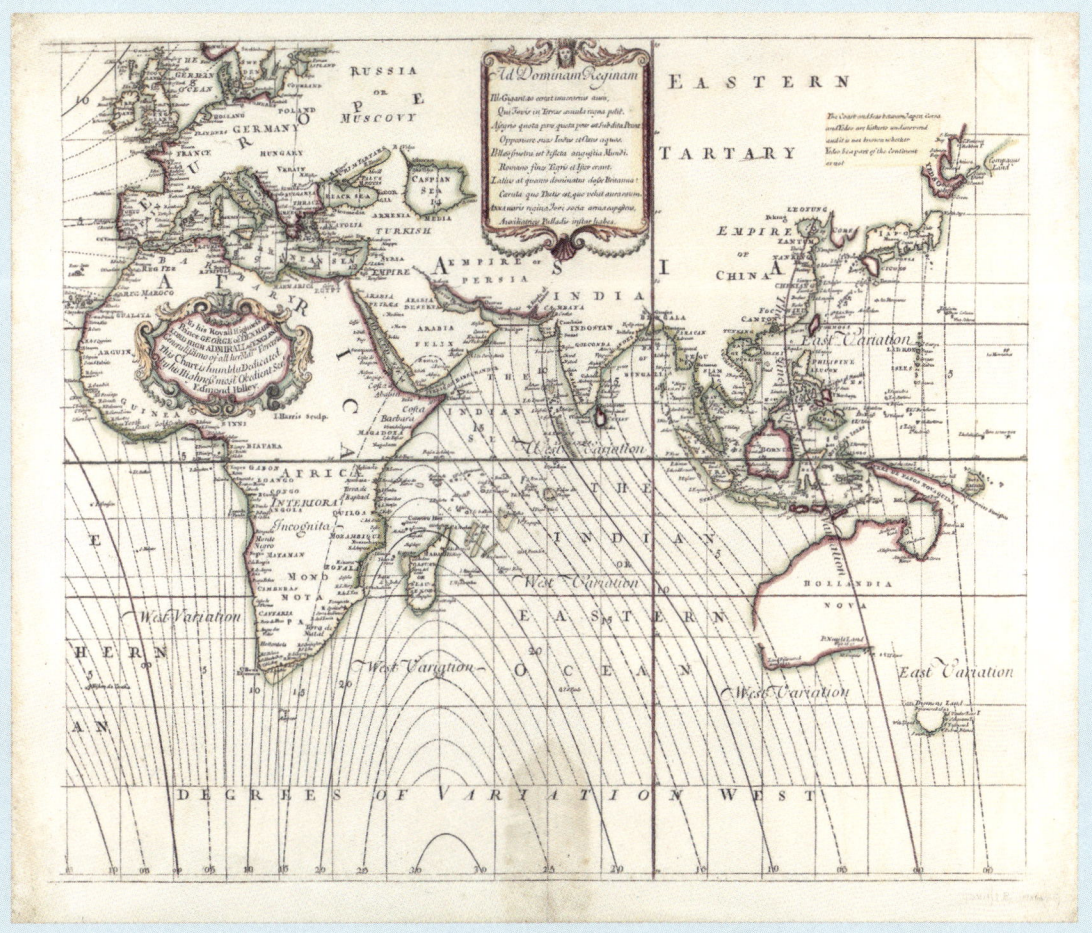

*A new and correct sea chart of the whole world shewing the variations of the compass as they were found in the year M.D.CC.*, Edmond Halley, published by Mount and Page, 1702
(G201:1/1B)

remained a lively topic of debate within academic communities. At the University of Cambridge (40), students and scholars – including the future Astronomer Royal, George Biddell Airy – were part of a movement to reform the physical sciences in Britain. The protagonists advocated the adoption of better Continental-style mathematical techniques, along with a broader range of subjects that included magnetism. In parallel, observatories across Europe were starting to broaden their range of so-called 'observatory sciences' to incorporate magnetic and meteorological observations.

By 1831, activists at Cambridge and elsewhere were dissatisfied with the state of science in Britain and created a new organisation called the British Association for the Advancement for Science (BAAS), which became a platform for radical ideas. One of the key speakers at the 1833 BAAS conference was Samuel Christie – father of the eighth Astronomer Royal, William Christie – who bemoaned the fact that Britain was the only country in Europe with no magnetic observations at its national observatory. A year later, German physicists Carl Friedrich Gauss and Wilhelm Weber created an international

Silver medal commemorating Carl Friedrich Gauss, Brehmer, 1777–1855 (MEC2666)

Gold snuff box, Alexander James Strachan, 1808 (ZBA0803)

network of magnetic observatories known as the 'Magnetische Verein' (Magnetic Union) to collect data on a global scale. This sparked interest among the Observatory's own Board of Visitors who argued for the resumption of magnetic observations at Greenwich, but it was becoming clear that the incumbent Astronomer Royal, John Pond, was too sickly to respond. By the time of his appointment as seventh Astronomer Royal in October 1835, Airy was already in correspondence with Gauss and preparing to instigate a new programme of magnetic observations that matched both his Cambridge-inspired ideas around scientific reform and his sense of duty in fulfilling the Observatory's obligation to international science.

## Practical motives

Airy's new regime of magnetic observations resonated with both the scientific ambitions of organisations such as the BAAS and the Royal Society, along with the practical and strategic interests of the Admiralty, the Observatory's funding body. On a national level, the BAAS had already started in 1833 to enlist people to undertake a magnetic survey of the British Isles, coordinated by Irish physicist Humphrey Lloyd and Royal Artillery Officer Edward Sabine at Woolwich Arsenal. On an international level, the Prussian explorer Alexander von Humboldt approached the Royal Society in April 1836 to enlist British support in extending the Magnetische Verein across its colonial observatories. The influential mathematician and astronomer John Herschel took up the case and collaborated with Gauss to establish new geomagnetic observatories alongside existing astronomical observatories within the British Empire. Herschel also secured funding and Admiralty support for a voyage to Antarctica to locate the south magnetic pole. With previous experience from similar efforts to find the north magnetic pole in 1831, James Clark Ross was the obvious choice to lead the voyage (1839–43) and it was a great success. As an Admiralty-funded place of routine observation with trained staff and high-quality instruments, the Observatory's new Magnet House (46) was the ideal scientific and national institution for providing comparative magnetic data with the Ross expedition. However, by the time Ross returned from his voyage in 1843, Airy had become

*Commander James Clark Ross*, John Robert Wildman, 1834 (BHC2981)

resentful of the workload associated with the magnetic observations and argued that they should cease. His protestations were overruled by his scientific peers and the work continued. A few years later, Airy's staffing concerns were mitigated by the introduction of photographic self-registering instruments that created an automatic and continuous record of both the magnetic and meteorological readings.

Somewhat ironically, the Observatory's magnetic data became even more important in the following decades as Victorian engineers encircled the globe with submarine cables (61). Understanding the Sun–Earth relationship now became a matter of practical rather than philosophical enquiry as telegraph operators experienced shocks and burns during periods of intense magnetic activity induced by the Sun (52, 53). Airy was also called upon by the General Steam Navigation Company to conduct experiments on the disruptive effect of new ironclad ships on compasses, for which he received a gold snuff box. Magnetism would remain an essential element of the Observatory's remit at Greenwich until 1925 when the encroaching electric railway lines made the observations impractical and the department relocated to Abinger in Surrey (91).

## 46
## THE MAGNET HOUSE

Keen to participate in the trend for magnetic research, Airy secured more land from Greenwich Park to install his new Magnetic Observatory just south of the Meridian Observatory. It was completed in May 1838 with the wooden cruciform structure extending 40ft (12m) across. The northern arm was partitioned off to create a computing room also used for resting between night observations. The remaining arms each contained a magnetometer (68), the scales of which were read by an observer using a theodolite telescope positioned in the centre of the building to avoid disturbing the sensitive instruments. Believed to have been painted either by Airy's wife Richarda or sister-in-law Elizabeth, this detailed watercolour shows two assistants at work: one is checking the ground thermometers for measuring temperatures at increasing depths, while the second climbs the 80ft (24m) electrometer mast to check the lantern for detecting atmospheric electricity. The small hut on the right, originally used during Captain FitzRoy's circumnavigation on HMS *Beagle* (1831–36) in which Charles Darwin famously participated, was later donated to Greenwich and housed a dip circle (54).

The assistants worked to a gruelling schedule with observations taken every two hours, and Airy soon realised he needed more staff. Thankfully, James Clark Ross's impending expedition to locate the south magnetic pole in 1839 helped Airy's cause. When the Admiralty asked him to provide comparative magnetic data for the expedition, he successfully negotiated extra funds for more observers. A few years later, Airy relieved staff from magnetic duties by installing an automatic photographic recording system, developed by London surgeon Charles Brooke, that created a continuous trace on light-sensitive paper. In addition, a roof platform was completed just in time for observations of the annual Perseid meteor shower in August 1872.

By the 1890s, the Magnetic Observatory was dwarfed by other buildings containing ferrous materials (77, 78) and the instruments were relocated to a new pavilion further away in Greenwich Park. The site of the former Magnet House is now occupied by the bronze cone of the Peter Harrison Planetarium.

➜
*The Royal Observatory, Greenwich, The Magnet House*, Richarda Airy or Elizabeth Smith, 17 July 1847
Cambridge University Library

47

## AIRY TRANSIT CIRCLE

Embedded within the Meridian Observatory, this telescope was designed by the Astronomer Royal himself and became the site's defining instrument for time and longitude. Drawing from his predecessors' work, Airy combined the functions of Bradley's transit telescope (17) with Pond's mural circle (35) to create an instrument bigger and more accurate than before. With an optical tube nearly 12ft long (3.7m) and 8.1in. (21cm) in diameter, the telescope was supported on two stone piers and kept aligned with the meridian by two smaller collimation telescopes positioned north and south. It became the defining instrument of the Prime Meridian of the World in 1884 (69).

Within a few years of installation, Airy added an automatic recording system to reduce human error. As each star crossed the field of view, the astronomer at the eyepiece in the pit underneath pressed an electric contact that pricked a hole into a paper-covered revolving drum. He could then time his observations afterwards by comparing the pinpricks to a parallel series of clock pulses sent via telegraph wires from the Hardy Regulator. Other colleagues in the Time Department upstairs converted this sidereal time into GMT, ready for national distribution via the Shepherd Motor Clock (49) and its telegraph network. For observations of the fast-moving Sun, a second observer was positioned at the circular array of microscopes on the western pier to read the finely engraved scales.

As the Observatory's main instrument, the Airy Transit Circle was serviced weekly and had its own accessories, including a trough filled with mercury for checking vertical alignment and a box of cobwebs for repairing broken wires within the eyepiece (79). Despite these efforts, the integrity of the observations was threatened by external factors. Tremors from the encroaching railway lines through Greenwich Park were a major concern during the 1860s, while the construction of Greenwich Power Station on the meridian in 1906 added more disruptive noise, smoke and vibrations. Nonetheless, the telescope was used to make over 600,000 observations during its century of service, with the last official observation made by Gilbert Satterthwaite in March 1954.

➔ The Airy Transit Circle, as shown in *The Midnight Sky*, Edwin Dunkin, 1869 (PBG3833)

The Airy Transit Circle (left) and its circular array of microscopes embedded in the western pier (above), Ransomes and May, Troughton and Simms, 1850 (AST0991)

# THE FIREBALL METEOR OF 1850

On the night of 11 February 1850, a young farmer in Oxfordshire was cutting hay when he found himself in a blaze of light. His first thought was that he had set the farm alight with his lantern but, as he turned, his eye 'was caught by a large ball of fire with a bluish light in the sky'. He was one of countless witnesses who saw this fiery visitor streak across the sky in a brilliant array of colours before disintegrating into a shower of sparks. Some even heard the delayed rumble of the final explosion that made dogs bark and furniture shake.

Over the next few days, accounts appeared in newspapers, sometimes accompanied by illustrations such as this evocative mezzotint by artist and sculptor Matthew Cotes Wyatt. At Greenwich, Airy recalled how he was distracted from his paperwork in the Computing Room by a 'brilliant body' that illuminated the sky as it traversed it like a rocket. The Superintendent of the Magnetic and Meteorological Department, James Glaisher (37), put out a nationwide call for observations and eventually published 45 accounts in a philosophical journal. From this data, he concluded the meteor first appeared above Shropshire and moved eastwards before exploding above Biggleswade, all within seconds. Unlike others who believed meteors were produced by lunar volcanoes or within Earth's atmosphere, Glaisher confidently stated: 'It seems certain that this meteor must have come from the regions of space far beyond the influence of our vapours.' Within 20 years, his hunch was vindicated as astronomers confirmed the extra-terrestrial origin of comets, meteors and fireballs.

Glaisher and colleagues began regular meteor observations at Greenwich in 1865 and a dedicated platform was installed on the roof of the Magnet House (46) in 1872. He resigned two years later, but observations of the Leonid meteor shower continued each November for another three decades.

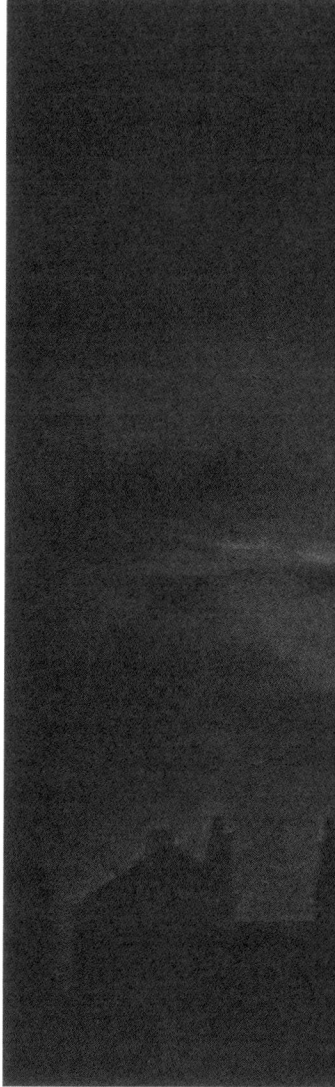

*A Representation of the Meteor Seen at Paddington about 12 Minutes before 11 o'clock, on the Evening of the 11th of Feb. 1850*, Matthew Cotes Wyatt, 1850
(PAJ3495)

## 49

## SHEPHERD MOTOR CLOCK AND GATE CLOCK

Having read about the successful use of telegraph networks to distribute time signals in the United States in 1848, Airy realised he could use the same technology to advance the Observatory's work. He discussed this with Charles Walker, telegraph engineer for the South Eastern Railway Company, and was impressed by demonstrations of Charles Shepherd's 'galvanic' (electric) clock at the Great Exhibition in 1851. Airy persuaded the Admiralty to pay for such a timekeeper and the Shepherd Motor Clock was installed in August 1852, along with the 3ft (92cm) diameter Gate Clock embedded in the outer wall – its mechanism accessible from the courtyard. The Motor Clock used a pendulum regulated by battery-powered electromagnets to produce a consistent swing. As it moved, electric contacts opened and closed, generating impulses that kept a series of wired 'sympathetic' dials in sync. The Shepherd Motor Clock became one of the Observatory's most important timekeepers, showing Greenwich Mean Time (GMT) as measured by observations with the Airy Transit Circle (47). Its connections and dials were used across the site for testing chronometers (51), dropping the time ball (39) and regulating the Gate Clock as a public display of time.

Thanks to Walker, the Observatory soon began distributing time signals to London Bridge station and beyond, reaching clocks and time balls across the nation. Airy also negotiated the installation of telegraph wires across Greenwich Park, connecting to submarine and overland cable networks (61, *Cable production in Greenwich*), fulfilling his ambition of sharing time signals to improve the known longitude of observatories around the globe.

The Shepherd Motor Clock remained in use until 1894 when it was superseded by the regulator Dent no.2012. The Gate Clock is now regulated by time signals from Global Positioning System (GPS) satellites and continues to show GMT year-round on a 24-hour dial, even, to the confusion of visitors, during British Summer Time (BST).

Motor Clock (left) and Sympathetic Dials, Gate Clock dial and movement (right), Charles Shepherd, 1852
(ZAA0531, ZAA0533)

## 50

## BRITISH LOCAL TIME MAP

At the Great Exhibition in 1851, while Airy was busy admiring Shepherd's clock (49), jeweller and watchmaker Henry Samuel Ellis was displaying his patented shawl brooches, five of which had already been purchased by Queen Victoria. It was a prestigious location in which to showcase his work and a great opportunity to explore his wider interests in science and engineering, especially the railways.

Alongside his business in Exeter, Ellis was an entrepreneur who created the Railway Shareholders' Association to persuade others to invest in the burgeoning technology. As Director of several Devonian railway companies, he tried to influence towns west of Bath to adopt GMT rather than local time, possibly using this bespoke map to promote his cause. Each vertical line represents a time difference of 1 minute east or west of Greenwich, with places already using GMT marked in bold. When the railway companies first emerged in the 1830s, they based their timetables on the local time of the main terminus. With a 30-minute difference between cities from east to west, it became increasingly difficult to coordinate timetables for long-distance journeys. In November 1847, the companies agreed to adopt GMT across the entire network, but some locations – especially those furthest east and west – were slow to change, as indicated in italics on the map. Even today, the clock at the Corn Exchange in Bristol has a pair of hands separated by 10 minutes, one for local time and one for GMT. It would take another three decades for GMT to become legal civil time across Britain in 1880.

In addition to his advocacy for GMT, Ellis was a keen amateur astronomer. He became a Fellow of the Royal Astronomical Society in 1855 and travelled to Spain with other wealthy amateurs and the Astronomer Royal to witness the total solar eclipse on 18 July 1860. Unfortunately, Ellis and his fellow observers in Santander saw very little under cloudy skies. The businessman returned to Exeter and used his wealth and local influence to create a free library, museum and science school for the city, becoming Mayor in 1868.

Diagram showing local times across Britain, Henry Ellis and Son, 1852
Cambridge University Library

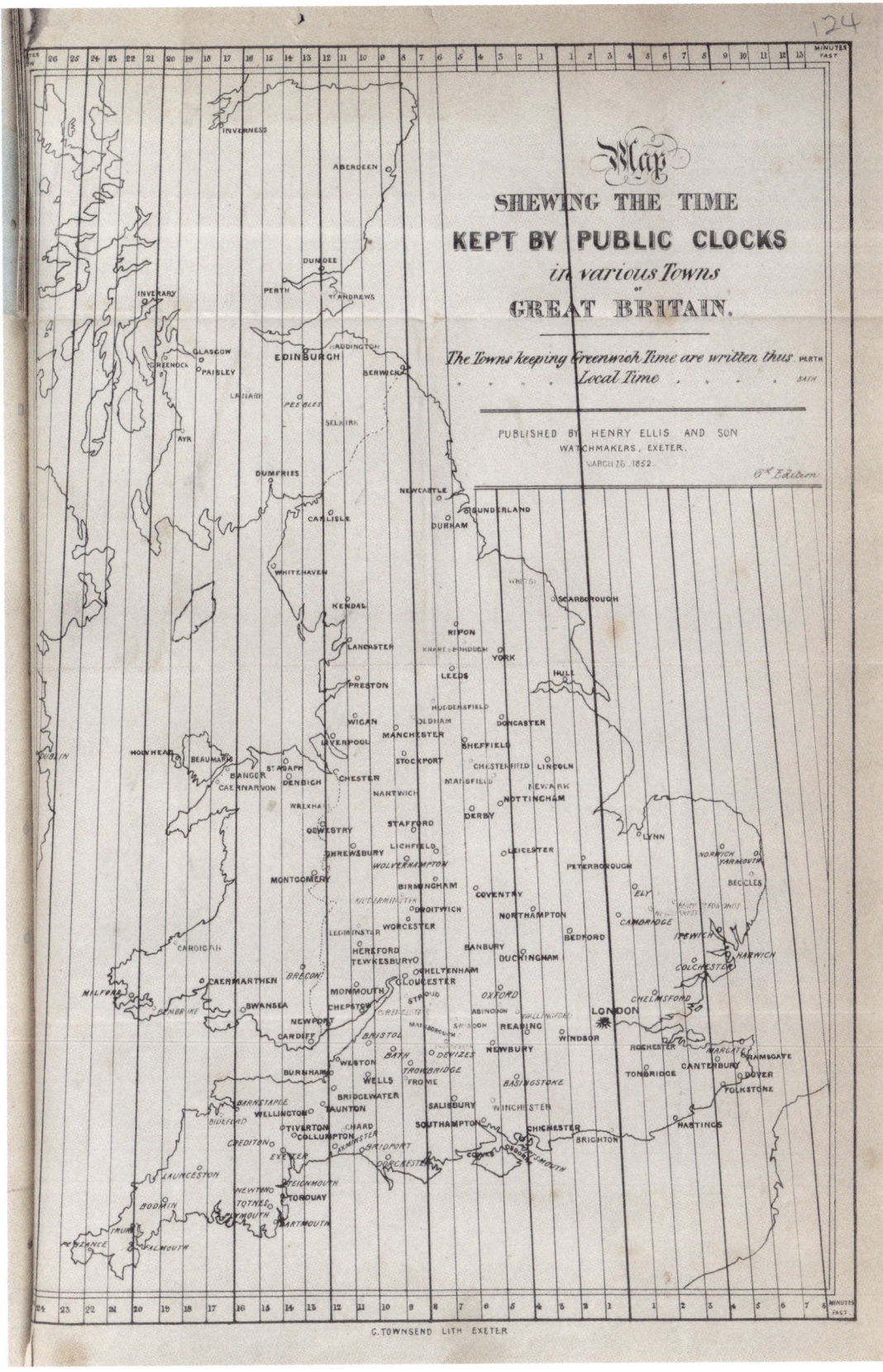

# MARINE CHRONOMETER Nº.10 BY BARRAUD

As shown by the last line on this Observatory ledger page, the chronometer known as 'Barraud 10' was returned to Greenwich on 3 August 1858 after nearly three years at sea. It was just one chapter in this instrument's astonishingly long career, which extended across three centuries from 1796 to 1967. With an upper dial for hours and minutes and a lower dial for seconds, this is a standard design chronometer set within a gimbal mount and mahogany box, complete with inlaid ivory nameplate that features the distinctive Admiralty broad-arrow mark. The accompanying ledger page starts in December 1821 when this one-day (wound daily) chronometer was issued to HMS *Barracouta* in preparation for a five-year expedition to survey the African coast. Each line documents the instrument's journeys across land and sea as it was transferred between ships, Admiralty locations and clockmakers for occasional repairs and servicing.

This instrument was just one of thousands of Admiralty-owned chronometers that were tested and rated against the Observatory's accurate clocks between voyages. After the initial Premium Trials of the 1820s (38), Airy suspended annual chronometer trials upon his arrival at the Observatory in 1836 but they were resumed in 1840 and became an essential part of the institution's daily routine, despite his protestations that the 'chronometer business' was a distraction from essential astronomical work. These delicate instruments were managed by assistant John Belville, who also established his own side hustle in sending a chronometer around London to sell accurate time to clockmakers (74). Every day for the year-long trial, Belville and his assistant checked around 30–70 chronometers – enduring the 'insect-like chatter of the ticking' – to record whether the instruments were running too fast or too slow (their 'rate'). To simulate conditions at sea, the instruments were subjected to variations in temperature either by placing them for a month outside on the northern wall or by placing on a tray near the stove; a purpose-built oven to simulate higher temperatures was later installed in 1869. Those chronometers that demonstrated the smallest amount of variation in rate were usually purchased by the Admiralty.

'Barraud 10', one-day marine chronometer, Howells, Barraud and Jamison, 1796 (ZBA0676)

Ledger entry for 'Barraud 10', p.145 (CSR/1)

## 52

## PUBLIC IMPERIAL STANDARDS OF LENGTH

Admired by millions of passers-by each year, these metal bars next to the Observatory's gates were installed in January 1859 to support its growing role as a place of public reference for standard length, time (49) and atmospheric pressure (57).

Two decades earlier, the nation's defining set of imperial units, the standard pound (0.45kg) and standard yard (0.91m) had been destroyed in the fire that swept through the Palace of Westminster in October 1834. As designs for the new Gothic-style building emerged in 1838, Airy was appointed Chair of the Parliamentary Commission responsible for reinstating the national standards. According to the Weights and Measures Act of 1824, the imperial yard was defined as the length of a pendulum with a one-second mean time swing at the latitude of London, corrected for use at sea level and in a vacuum. But by the 1830s, scientists were less convinced about using pendulums for this definition because the corrections for altitude and atmospheric conditions had proved insufficient.

Instead, Airy and the Commission opted to replicate the old imperial yard with a new metal measure. To find the best material, they conducted experiments immersing metal bars in hot water and troughs of mercury across a range of temperatures. They eventually settled on a solid, square-shaped bar made from a gun-metal alloy of copper, zinc and tin for minimum distortion and maximum durability. Each bar was 38in. (96cm) long with gold plugs exactly 1 yard or 36in. (91cm) apart, as measured at 16.7°C (62°F).

Reference copies were made in the 1840s for the Palace of Westminster, the Royal Society and the Royal Mint. Airy stored the Observatory's versions in the Record Room and returned them to the Board of Trade every ten years for validation. These standards are now stored in the Museum's collections. Other public standards can still be found in London's Trafalgar Square and at the Guildhall in the City of London.

Public standards of length, installed within the Observatory's exterior wall, Troughton and Simms, 1859 (AST1048)

Standard measures and public barometer on the front wall of the Royal Observatory, Greenwich, unknown photographer, about 1930 (AST1107.20)

## 53
## MAGNETOGRAM OF THE CARRINGTON EVENT ON 1 SEPTEMBER 1859

On Thursday 1 September 1859, amateur astronomer Richard Carrington undertook his daily routine of observing the Sun from his private observatory at Furze Hill, Surrey, about 20 miles (32km) from Greenwich. Carrington had started work on a catalogue of 3,753 stars but later switched to solar observations. Frustrated by the lack of interest shown in the subject by national observatories such as Greenwich, he decided to record sunspots over the next 11-year solar cycle. Using his 4.5in. (11.4cm) equatorial telescope by Troughton and Simms, he projected the solar disc onto a coloured glass plate and sketched the daily sunspot patterns using ultra-thin gold wires to plot their position.

At 11.18 a.m. that sunny morning, Carrington was sketching a particularly large complex of sunspots when two intense areas of white light appeared, slowly morphing into kidney-shaped patches. He rushed back to the house to seek a witness but within minutes the lights had disappeared. Twelve hours later, aurorae lit up the atmosphere as far south as the Caribbean, with similar effects seen in the southern hemisphere. People marvelled at reading newspapers in the glow and telegraph operators experienced disrupted transmissions and electric shocks from sparking apparatus.

The association between solar activity, aurorae and magnetic needle motion had been speculated upon since the 1740s but astronomers now had better technology to investigate. At Greenwich and other magnetic observatories, the recording drums alongside the magnetometers (68) captured intense needle movements, generating jagged magnetogram peaks for days.

Despite seeing the Kew Observatory magnetograms, Carrington was reluctant to claim a causal link between solar activity and terrestrial effects, stating that 'one swallow does not make a summer'. Nonetheless, his work proved the value of sunspot observations and set the trend for solar studies at Greenwich in later decades (64, 72).

Photographic record of magnetometer measurements for 1 September 1859 at the Royal Observatory, Greenwich
The British Geological Survey

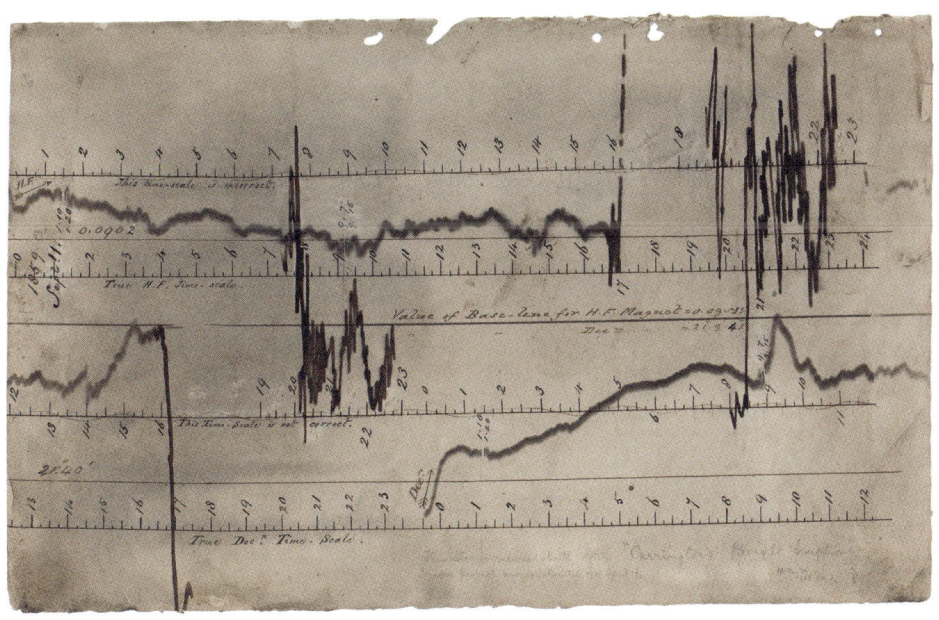

## 54

## AIRY'S DIP CIRCLE

While people marvelled at the dramatic effects of the Carrington event (53), Airy was busy finishing the design of his own magnetic instrument. The changing vertical angle between the tip of a compass needle and the horizon, known as 'dip' or 'inclination', had been recognised by navigators and instrument makers since the sixteenth century, along with its horizontal counterpart (see *The Age of Magnetism*). At the equator, the needle of a dip circle lies flat (0°) but at the magnetic poles it turns vertically (90°). On 1 June 1831, naval explorer James Clark Ross used a dip circle to locate the north magnetic pole at Boothia Peninsula, northern Canada. As more data was collated by explorers and observatories, it became apparent that dip instruments could be used to map the gradual drift of the Earth's north magnetic pole, a trend that continues today.

For Airy, changes in the Observatory's own dip results were proving problematic. In the winter of 1856 he received a letter from Professor Christopher Hansteen of Christiania (Oslo) University who was concerned by the anomalous dip results between Greenwich and other observatories in northern Europe. Airy duly consulted Edward Sabine at Kew Observatory for some comparative data and agreed with Hansteen's analysis. He then carefully examined the Observatory's dipping needle, made by the London maker Thomas Charles Robinson, and realised that the needle was positioned incorrectly, causing it to catch on the brass scale. Despite adjusting the instrument and following Hansteen's advice to obtain better needles, systematic errors still appeared in the results, leading Airy to admit ruefully in his Annual Report for 1859, 'In regard to the Dipping Needles, I am still very much perplexed.'

Responding in his typical fashion, thereafter Airy set about designing his own instrument, creating a dip circle that featured a more stable base, microscopes for reading the needle movements more closely and a revolving gaslight that illuminated the scales for both daytime and night-time observations. Standing 25in. (64cm) tall with a highly stable weight of 135lb (61kg), the instrument was eventually installed in October 1861 and, after a few more years of adjustment, remained in use until December 1914.

Airy's dip circle, Troughton and Simms, 1860 (AST0701)

Detailed view of the dipping needle and the makers' signature

146 A HISTORY IN OBJECTS

## 55
## HOURLY TIME SIGNAL RELAY

Situated in the Time Department under the supervision of William Ellis, this relay device was the nerve centre of the Observatory's time distribution system from 1861 to 1883. Every hour, mean time signals from the Shepherd Motor Clock (49) activated the relay's electromagnets, closing the contacts for six telegraph wires extending across London and beyond to the time ball at Deal on the Kent coast. Signals were transmitted hourly, day and night, to the General Post Office and the London and South Eastern Railway Company at London Bridge. These were then redistributed twice daily, at 10 a.m. and 1 p.m., to post offices and stations nationwide. Some signals triggered local time devices such as electric bells, time balls and even guns, as heard in Edinburgh and Newcastle upon Tyne. Signals also went to the British Horological Institute in Clerkenwell for chronometer makers, though many still relied on the traditional service operated by the Belville family (74).

Designed by London engineers William Morris and Henry Mapple, this relay helped extend and improve the reliability of the Observatory's time service, now an essential part of Victorian daily life. In his Annual Report for 1853, Airy proudly wrote, 'I cannot but feel a satisfaction in thinking that the Royal Observatory is thus quietly contributing to the punctuality of business through a large portion of this busy country.'

But, as with any system, there were challenges. Interruptions occurred periodically, especially in winter, when aerial telegraph cables were vulnerable to wind and snow. On average, the Deal time ball failed to drop about once a year; in such cases, a black flag was hoisted to warn mariners and the ball was released at 2 p.m. instead.

Hourly time signal relay, William Morris and Henry Mapple, around 1861 (ZAA0718)

148 A HISTORY IN OBJECTS

## 56
## SINGLE-NEEDLE TELEGRAPH

At first, telegraph offices were based at major railway stations and only wealthy business leaders could afford the cost of sending messages. As the telegraph network expanded, demand grew and by the late 1860s, five companies in Britain were carrying 6.8 million messages a year, over 91,000 miles (146,450km) of wire.

This single-needle telegraph machine is typical of those used in the 1860s by operators who relied on the Observatory's time signals. Letters were generated by moving the handle left or right, forming Morse code sequences: left turns for dots, right for dashes. Common letters like 'a' and 'e' were quick to transmit; rarer ones needed longer sequences. At 10 a.m. and 1 p.m. daily, operators cleared the line to await the GMT time signal from the Shepherd Motor Clock (49) and relay (55), ensuring accurate message timing. Telegraph machines also became tools for scientific study, as operators observed strange effects – sparks, violent needle movements – linked to electric currents in the Earth (58) and sunspot activity (53).

The Observatory's influence on telegraphy extended worldwide. In 1855, Airy's galvanic superintendent Charles Todd left Greenwich to lead the construction of a telegraph line across Australia. Over the next two decades, he created a chain of stations, one forming the nucleus of a settlement later named 'Alice Springs' after his wife. Completed in August 1872, the Overland Telegraph Line enabled messages to be sent between England, India, Singapore and Australia in hours rather than months. It also supported longitude measurements and time signal distribution for the 1874 transit of Venus expeditions (64, 65).

➔
Single-needle telegraph machine, unknown maker, about 1860–70
(ZAA0656)

IN FOCUS

# Meteorology becomes a science

The commencement of official, comprehensive weather readings at the Observatory in 1840 reflected changing scientific and public attitudes to meteorology, the scientific study of the atmosphere that stems from the Greek word *meteoras* meaning 'lofty' or 'in the air'. The study of astronomy was centuries old, but this new discipline took time to gain acceptance. In early nineteenth-century Britain, observing and understanding the weather was the preserve of those whose livelihoods revolved around seasonal weather patterns, such as sailors, fishermen, farmers and gamekeepers. Their knowledge was based on folklore and lived experience that had been handed down the generations, including the well-known saying 'red sky at night, shepherds' delight' as an indication of fine weather on the day after a brilliant sunset.

In the previous centuries, meteorological readings at Greenwich had only been taken sporadically and for a specific purpose, namely to see if localised atmospheric effects could affect instruments and observations. Flamsteed and his successors sometimes noted the readings of nearby barometers and thermometers alongside their astronomical results but did not perceive the data to be useful to anyone else. By the 1820s, academic interest in meteorology was gaining ground within universities, as scholars included the subject within the wider endeavour of measuring, classifying and understanding natural phenomena. Just like the emerging studies of terrestrial magnetism (see *The Age of Magnetism*) natural philosophers began to interpret and describe atmospheric effects in terms of mathematical laws, physical causes (heat, motion) and the chemical properties of matter. With their pre-existing infrastructure of instruments and trained observers, national and private observatories became the obvious players in this new global project to collect meteorological data.

**Early meteorological observations at Greenwich**
The first systematic meteorological observations at Greenwich began in 1807 when the fifth Astronomer Royal, Nevil Maskelyne, started to make daily observations. His successor John Pond expanded the range of observations to include rainfall (37) from December 1814 and later added cloud cover, wind direction and speed to create a more comprehensive record. Pond's assistant John Belville (51, 74) also kept a private daily meteorological journal as he moved between various homes in Greenwich and nearby Blackheath across four decades. Additional meteorological observations were made at the Observatory in 1839 when George Biddell Airy agreed to participate in Sir John Herschel's scheme to collect data from various observatories on the equinoxes and solstices (around 21 March, June, September and December). Herschel was an influential figure in the scientific community and his involvement with the Observatory's Board of Visitors presumably helped persuade Airy to allocate staff time to this new task. Once Airy had created the new Magnetic and Meteorological Department in November 1840, the observations were overseen by Superintendent James Glaisher (37).

**Making meteorology useful for the nation**
Unlike previous astronomers at Greenwich, Glaisher was keen to make meteorological data available and useful to a public audience. In 1847 he used his contacts among 'gentlemen of education and position' to create a national network of around 60 volunteer observers from

*Diagram of Meteorology*, James Reynolds, 1846 (AST0051.7)

Dundee to Plymouth, who collected daily weather readings from a prescribed set of instruments. The results were returned to Glaisher via telegrams that he used to provide reports (Meteorological Returns) on the previous day's readings to the *Daily News*, along with a comment on the monthly averages. He also recruited railway stationmasters to complete and send daily readings to London via the first available train. These preprinted forms were collected by a messenger at midnight to be published in the morning edition of the *Daily News*. The value of this telegraphic system became apparent a few years later when Glaisher compiled the incoming data into a weather map that was sold at the Great Exhibition every day except Sunday from August to October 1851. A coded set of symbols alongside each named city provided information on the wind direction, barometer reading and overall status: fine, cloudy or rainy.

In addition to his newspaper reports, Glaisher was called upon to advise on potential links between the weather and public health. In 1855 the Greenwich astronomer submitted a report to the Houses of Parliament 'on the meteorology of London and its relation to the epidemic of cholera', in which he surmised that the outbreaks in previous years had been exacerbated by high atmospheric pressure and thick mists that kept the stagnant, disease-ridden air close to residents. Although this 'bad air' (miasma) theory of infection has since been displaced by germ theory, it is an interesting example of how meteorologists tried to promote the scientific nature and public utility of their work. Similarly, in 1863 Glaisher collected and assessed meteorological data from independent observers located across India to consider how British troops might cope with the unfamiliar weather conditions, concluding that 'the climate of the hill stations in India is dry enough, the temperature low enough, and the sky clear enough, without any excess rain for Europeans'.

## Meteorology is established

Ten years after the creation of the Magnetic and Meteorological Department at Greenwich, the British Meteorological Society was founded by a group of gentlemen on Wednesday 3 April 1850 at Hartwell House, the stately home of astronomer Dr John Lee. Glaisher was appointed as Secretary and the new Society – awarded royal status in 1883 – helped consolidate previous embryonic meteorological societies into one group. There was growing recognition of the subject's scientific merit in government too with the creation of Meteorological Department at the Board of Trade in 1854 – the origins of today's Met Office – that was led by Captain Robert FitzRoy. A few years later, the Great Storm of October 1859 revealed how understanding the weather could potentially save lives at sea. Over 133 ships were lost during the stormy conditions on 25–26 October 1859, including the steam clipper *Royal Charter*, which was smashed apart on the rocky coast of Anglesey, Wales. The event gave FitzRoy the impetus and funding to set up a national storm warning service for coastal communities that began in February 1861. Once FitzRoy had received the daily weather reports from his network of observers, he assessed the data and then issued warning telegrams to the locations most at risk. If they received such a warning, the harbourmasters would hoist a coded signal composed of cones and flags to alert mariners. Similarly, Glaisher worked with the Royal National Lifeboat Institution (known today as the RNLI) to provide barometers at lifeboat stations (57) so that crews could see first-hand any changes in local atmospheric conditions that might foretell approaching storms. Meteorology was no longer just a scientific curiosity but a matter of life and death.

Despite the increasingly robust mathematical and scientific nature of meteorology, the Astronomer Royal remained unconvinced by the lack of a theoretical framework for the subject and voiced his scepticism in a letter written in 1861 to his Chief Assistant, Robert Main: 'For Meteorology, *nomine Scientia*, I have the most complete contempt. I set no value on it except as a kind of local statistics...'

## Data collection at Greenwich

Glaisher and his assistants relied on a suite of instruments across the site to complete their meteorological readings. Various barometers and thermometers were read every two hours while the rain gauges and anemometers for wind speed were checked once a day. The observers also recorded their visual estimations of wind speed and cloud cover, sometimes using the appearance of the Moon to gauge cloud thickness. Most of the instruments were clustered around the Magnet House but others were located elsewhere on the site, such as the anemometer for measuring wind speed and direction, situated within the north-west turret of Flamsteed House. Rain gauges were positioned on various roofs, while the thermometers were housed within wooden screens designed by Glaisher. In 1847, a new, innovative photographic recording system designed by the London surgeon Charles Brooke helped to alleviate Airy's concerns about

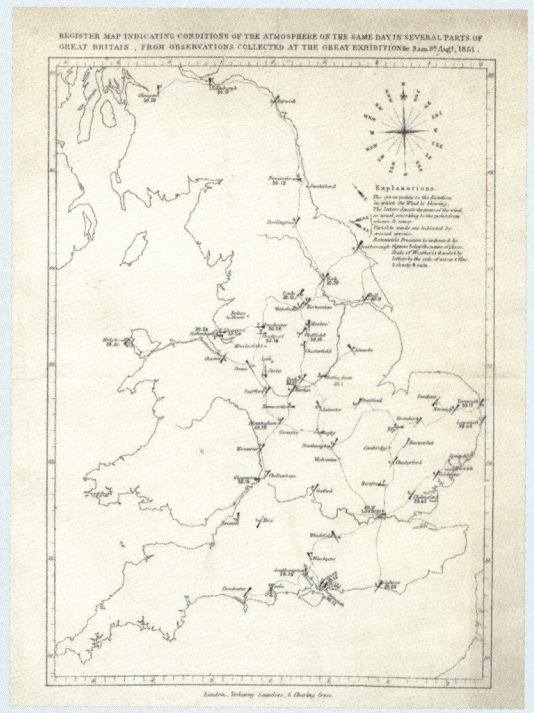

*Atmospheric Maps [...] from observations collected by the Electric Telegraph Company at the Great Exhibition, August [8th] to October 11th*, Royal Commissioners for the Exhibition 1851, Electric Telegraph Co. Trelawny Saunders, 1851
National Meteorological Library and Archive

*Loss of the 'Royal Charter'*, unknown artist, 1855 (PAH0291)

staffing by collecting the data automatically. Some of the meteorological instruments were later moved to a platform installed on the roof of the Magnet House to reduce the buffering effect of nearby trees. During the 1860s, Glaisher undertook additional meteorological readings from a suite of instruments that was installed within various hot-air balloons piloted by wealthy gentlemen 'aeronauts'. Over the course of 28 flights in four years, Glaisher meticulously recorded how instruments varied with altitude; he even narrowly escaped death when he lost consciousness for 30 minutes at an altitude of 6 miles (9.7km) during a flight in September 1862.

For Glaisher's staff and their successors, the daily meteorological readings became an essential part of the Observatory's routine that continued through both wars before ceasing in the early 1950s as other instruments were installed at the Observatory's new location at Herstmonceux (see *Epilogue*). A limited number of meteorological observations continued to be taken by Museum staff around the Observatory and from nearby sites for a few more decades, but these ceased in 2007. Today we only use rooftop anemometers to assess wind speeds in advance of events such as dropping the time ball and opening the telescope domes.

## PUBLIC BAROMETER

In 1857, Chief Meteorologist Rear-Admiral FitzRoy persuaded the Board of Trade to supply barometers to 40 fishing villages to reduce crew losses in stormy seas. In the following years, additional barometers were installed at other fishing villages and stations of the National Lifeboat Institution (now the RNLI). These instruments were made by the London firm of Negretti and Zambra and each one was examined and tested for accuracy by the Royal Observatory's Superintendent of Meteorology, as stated on the finishing label: 'This barometer reads correctly with Greenwich Standard, James Glaisher, FRS' (37, 48).

Despite continuing scepticism about the scientific basis of the subject, Airy recognised that meteorology was now an essential part of the Observatory's national service. In response to the growing demand for publicly visible barometers – and perhaps with some jealousy towards Glaisher's public profile – Airy decided to install such an instrument at Greenwich. The Observatory's daily readings of atmospheric pressure were already made using the 'Greenwich Standard', and a barometer made by John Newman that was located within the Magnet House (46). For the public instrument, Airy opted to work directly with Henry Negretti and Joseph Zambra himself and began to correspond with the Anglo-Italian duo at their Hatton Garden workshop in April 1862, filling his letters with design sketches.

The instrument was finally installed on 2 June 1864 and was situated on the outer wall alongside the Public Standards of Length (52). Airy proudly reported to the Board of Visitors that the barometer was 'a subject of great interest, and gives valuable information, not only to the public but even to ourselves'. It consisted of a 3.28ft (1m) long column of mercury that was embedded within the Observatory's outer wall. As the liquid metal rose and fell in response to changes in atmospheric pressure, the indicators on the brass dial showed the corresponding height in inches as 'Present' and 'Lowest' since 9 p.m. the day before. A magnet was used to reset the steel indicators. Apart from a few years' interruption as a result of bomb damage in October 1941, the barometer remained on continuous public view until it was removed in 1960.

Dial plate of the wall-mounted public barometer, Negretti and Zambra, 1864 (ZBA4519)

## 58

## EARTH CURRENT GALVANOMETER

While the Observatory's public barometer (57) offered a direct view of scientific measurements, other instruments on the site were busily recording invisible effects within the Earth. As the Victorian enthusiasm for communications spurred a rapid growth in telegraphy (56), operators began to notice occasional but irksome disturbances within the apparatus (53). In the early 1860s, Airy installed two electric cables to detect these 'spontaneous terrestrial galvanic currents'. He relied on the assistance of Charles Walker, telegraph engineer for the local railway company, to install a wire from the Magnetic House to Greenwich station and along the network to Croydon, while a second wire terminated in Dartford.

Each wire was read using a galvanometer. The instrument consisted of a sensitive magnet, suspended from the overhanging arm by a long thread, which was positioned between two copper coils. When the Earth's magnetic field induced a current in the telegraph wires, the galvanometer's coils became magnetised and the central needle moved in response. The greater the earth current, the greater the movement. In March 1865 Airy automated the system by adding a rotating drum of photosensitive paper that recorded a light trace reflected by a mirror located below the coil. In November 1881, his successor William Christie commented how 'during [a] great magnetic disturbance, the earth current motions were so violent that the records could not be traced'.

These sensitive wires and instruments required much maintenance over the years. In 1868, Airy opted for new cables that extended from Angerstein Wharf to Ladywell station, and from Lewisham station to Morden College. While improved, the new cables were still susceptible to failure, either through storm damage or from sabotage by metal thieves. By the 1890s, however, the greatest threat came from the railway itself, with electric trains generating too much disturbance for the instruments to detect any natural effects. The Observatory ceased its Earth current measurements at the end of 1909.

Earth current galvanometer, James White, 1864 (AST0702)

Detailed view of the coils

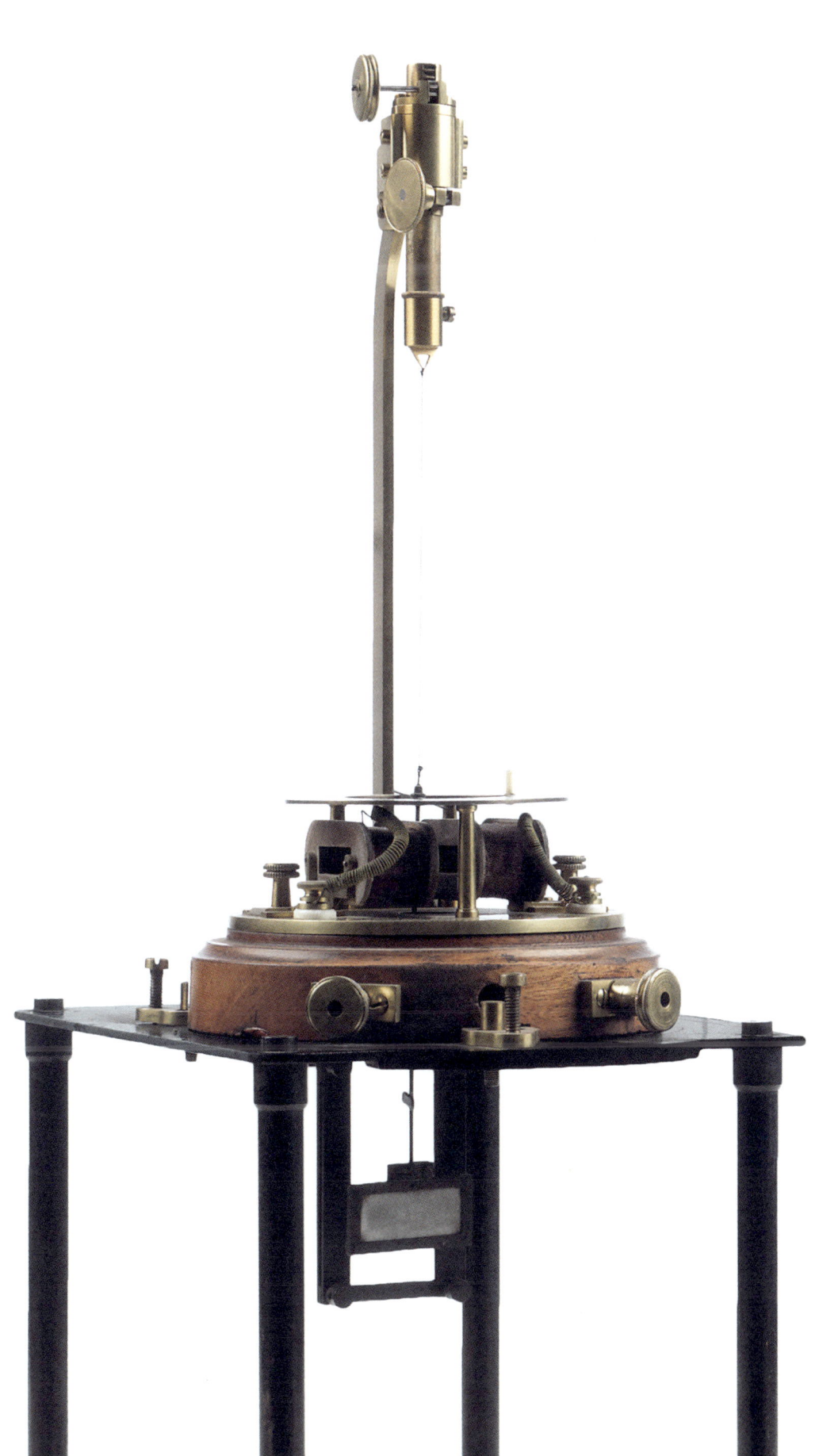

## 59
## PHOTO OF THE AIRY FAMILY

This fashionable scene reminds us of how the Observatory was both a place of work and a family home. The first three Airy children (George, Arthur and Elizabeth) were born at Cambridge Observatory (40) but the remaining six were born at Flamsteed House and survived into adulthood. Like other middle-class Victorian families, the boys were sent to the local grammar school in Blackheath, while the girls were educated at home.

The busy household was sustained by three domestic servants and a gardener who tended the fruit trees and kitchen garden. The Astronomer Royal kept to a strict daily routine: work from 9 a.m. to 2.30 p.m. followed by a brisk walk before dinner at 3.30 p.m., some rest, tea, more work until 10 p.m. and then an hour of reading or cards before retiring at 11 p.m. The family took regular breaks at their country cottage in Suffolk (41), while energetic holidays to the mountainous regions of Britain and Europe sustained their love of hiking. This spirit of adventure seemingly continued at home with Wilfrid and Hubert receiving a stern letter of reprimand from the Chief Assistant at Greenwich, Robert Main, for running around the roof of the Meridian Observatory with its dangerous open hatches above the telescopes.

While none of the Airy children became astronomers, their careers reflected the scientific influence of their unusual childhood home. Wilfrid and Hubert embarked on technical careers in civil engineering and medicine, respectively, and Osmund became an Inspector of Schools. Hilda married the Cambridge mathematician Edward John Routh, but Christabel and Annot remained at Greenwich. In 1870, their mother Richarda suffered a devastating paralytic stroke, prompting the two sisters to care for her until her death in 1875. They continued their caring responsibilities for their father when he retired to a nearby house in August 1881, remaining there until his death in January 1892.

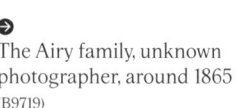
The Airy family, unknown photographer, around 1865
(B9719)

Back row, left to right:
Richarda, George, Hilda
Front row, left to right:
Annot, Osmund,
Christabel, Hubert, Wilfrid

## 60

## THE MIDNIGHT SKY (1869)

Just like well-known media figures today, astronomers at Greenwich have long sought to share their passion for astronomy with popular audiences. In 1861, Assistant Edwin Dunkin began publishing a monthly guide to the night sky in *The Leisure Hour*, a weekly magazine by the Religious Tract Society. With its mix of richly illustrated articles, *The Leisure Hour* was one of many Victorian periodicals offering instructive but entertaining content with a strong Christian ethos for a broad readership across classes, genders and ages.

Encouraged by complimentary letters from readers, Dunkin compiled his guides into a single volume called *The Midnight Sky* in October 1869. He enlisted his 20-year-old son Edwin as draughtsman to create star charts showing the constellations above familiar London landmarks like St Paul's Cathedral and the Royal Naval College at Greenwich. He also included charts for the southern hemisphere, using the profile of Table Mountain, Cape Town (45), as a reference point.

The second half of the book explored the Sun, Moon, planets, comets and meteors, followed by a more advanced section on the magnitudes, colours and proper motion of the stars. Dunkin kept the book current by adding new innovations such as astronomical photography and spectroscopy (67, see *Splitting starlight with spectroscopy*) to analyse stellar composition. As the book was revised, this section on 'new astronomy' expanded with technological and scientific developments.

Reviewers praised Dunkin's book, noting that 'it tends to shorten long evenings and give an increase of delight to whoever strolls beneath a star-lit sky'. He retired from the Observatory in August 1884 after 46 years' service but continued writing for *The Leisure Hour* until 1891.

⬇⬊
View of the January stars above London and frontispiece from *The Midnight Sky*, Edwin Dunkin, 1869 (PBG3833)

IN FOCUS

# Cable production in Greenwich

By the 1850s, the riverside wharfs of north Greenwich were home to the production of submarine cables that were transforming telegraphy from a national web of overland wires into a global submarine communications network. Situated over a mile (1.6km) away in Greenwich Park, the Royal Observatory was quick to become a beneficiary of this engineering marvel. The Astronomer Royal, George Biddell Airy, recognised its potential both for participating in collaborative scientific projects across the globe and for extending his institution's reputation and influence. Although there was little direct contact between the Observatory and its cable production neighbours, it was the beginning of a technological partnership that shaped the institution's work for decades. Submarine cable equipment is still produced in Greenwich to transmit internet data and time signals.

**Early submarine cable attempts in the 1850s**
Despite its brief success that only lasted a few days, the installation of an insulated copper wire from Dover to Calais in August 1850 demonstrated the feasibility of using submarine cables to transmit signals between continents. A year later, the link was reinstated with a better cable and messages began to flow between Paris and London. Buoyed by this success, businessmen began to turn their attention towards longer connections, leading to the creation of the Atlantic Telegraph Company in August 1856. A few months later, the investors placed an order with the Gutta Percha Company in Islington to produce 2,500 nautical miles (4,630km) of insulated copper wire as the essential core of the first transatlantic cable. The company was named after its key material, which was a natural type of rubber derived from the sap of the *Palaquium gutta* tree. In the early 1840s, the London experimentalist Michael Faraday had identified gutta percha's excellent electrical insulation properties, leading cable manufacturers to invest in its use. Manufacturers subsequently appreciated the material's lack of corrosion in salty seawater and its excellent performance in the high pressure and cold temperatures of the ocean depths. Supplies were limited as *Palaquium gutta* trees only grew in specific parts of the Malay Peninsula and surrounding areas. At the time, these regions were under British colonial rule so cable companies in Greenwich and elsewhere had a distinct advantage over their European rivals.

The first stage in the production process was to create a core composed of a single copper wire surrounded by six strands, followed by three protective layers of gutta percha, hemp and tar, held together by a sticky mixture that was developed and patented in 1859 as 'Chatterton's Compound'. Once complete, the insulated copper wire required covering with a protective sheath made from twisted iron strands – 'armouring' – a task that was divided among two companies: R.S. Newall and Company in Birkenhead, and Glass, Elliot and Company in Greenwich. Initially at Morden Wharf, and later Enderby Wharf, the Greenwich-based company took advantage of the large-scale facilities and riverside access that had once supported the manufacture of hemp rope. The workers at Glass, Elliot and Company started to armour their assigned half of the cable – around 1,250 nautical miles (2,315km) – and produced around 30 miles (48km) of cable per day. Once completed, the cable was coiled into iron tanks for transfer to the cable-laying ship.

Deciding when and where to lay the submarine cable was an equally complex process. In the early 1850s, the US Navy had undertaken a series of deep-sea surveying voyages that had revealed a relatively short and flat route across the

↑
*The reels of gutta percha covered conducting wire conveyed into tanks at the works at Greenwich*, Robert Dudley, F. Jones and Day & Son Ltd, 1865
(PAG8287)

Atlantic seabed from Newfoundland to Ireland. Hydrographer Lieutenant Maury concluded that a transatlantic cable was indeed feasible, although he was realistic about 'the possibility of finding a time calm enough, the sea smooth enough, a wire long enough, a ship big enough, to lay a coil of wire 1,600 miles [2,575km] in length'.

A few years later, in the summer of 1857, Maury's prediction came to fruition with the laying of the first cable, although the venture had to be aborted when the cable snapped around 280 miles from the Irish coast. Undeterred, the investors ordered another cable section and the transatlantic link was successfully completed in August 1858, a feat that was marked with celebratory messages exchanged between Queen Victoria and President Buchanan. Sadly, the celebrations were short-lived as the cable completely failed just two months later.

### Further attempts in the 1860s
Unsurprisingly, investors were initially reluctant to put more money into yet another expensive cable attempt but they were persuaded by a positive report in 1861 from a joint committee formed of the British Board of Trade and the Atlantic Telegraph Company – based on the evidence of over 40 expert witnesses – which provided the reassurance that another attempt was worthwhile. Greenwich cable manufacturer Richard Glass realised that he could only ensure the technical success of future cables by combining all elements of production, prompting him to work with investors to merge Glass, Elliot and Company with the Gutta Percha Company to create The Telegraph Construction and Maintenance Company (Telcon) in April 1864. Within a month it had secured the contract to produce a new transatlantic cable and began negotiations to hire the locally built steamship SS *Great Eastern* – the largest ship in the world – as its cable-laying vessel. Two smaller ships were used to ferry the completed cable from Greenwich to the deeper waters of Sheerness where the SS *Great Eastern* was eventually laden with over 2,490 nautical miles (4,611km) of cable.

'Receiving messages from the SS Great Eastern at the telegraph house at Valentia, western Ireland', *Illustrated London News*, 7 August 1865, p.117 (ILN/1865/47)

*The route of the Atlantic Telegraph shown with a section of 'The Great Eastern'*, Captain H. Clark and Stevens, 1865 (PAG8266)

*Interior of one of the tanks on board 'The Great Eastern'... Cable passing out*, Robert Dudley, McCulloch and Day & Son Ltd, 1865 (PAG8268)

With intense public interest in the technological marvel of the age, newspapers and publishers sought to capitalise on the event with special articles and volumes. One such company was Day & Son, which recruited the reporter William H. Russell and the artist Robert C. Dudley to join the 1865 expedition with a brief to document the entire cable-laying process. The resultant book, *The Atlantic Telegraph*, appeared shortly afterwards and was lavishly illustrated with 24 lithographs of Dudley's watercolours that vividly portrayed this momentous achievement.

Ten days after the ship's departure from Valentia in western Ireland, the cable broke. Despite repeated attempts to retrieve the cable using a grappling hook, the crew was unsuccessful and had no choice but to return home. On the next attempt in the summer of 1866, the crew of the SS *Great Eastern* not only completed the transatlantic link but also successfully located, retrieved and repaired the 1865 cable, thanks to a set of coordinates based on noon sights with a sextant (27), chronometer readings (51) and the sending of GMT time signals along the cable to plot the changing position of the ship. Britain and the United States were finally connected by two cables that could transmit around eight words per minute.

**The Observatory's use of submarine cables**
New submarine cables across the English Channel and the Atlantic offered innovative opportunities for astronomers to measure the difference in longitude between observatories. Once accurately defined, each observatory could then be used as a data point for the creation of new maps and charts. Before the era of cables, astronomers had physically transported crates of around 30 chronometers between observatories to determine the difference in local time and longitude. In 1844, American astronomers started to experiment with sending telegraphic time signals to measure the time difference between the transits of certain stars across each local meridian. Engineer John Locke automated this process with the invention of the chronograph – a rotating paper-covered drum that recorded the timings – which was installed at the US Naval Observatory in 1849. Airy was much impressed by the so-called 'American method' and decided to create his own telegraphic programme at Greenwich. In May

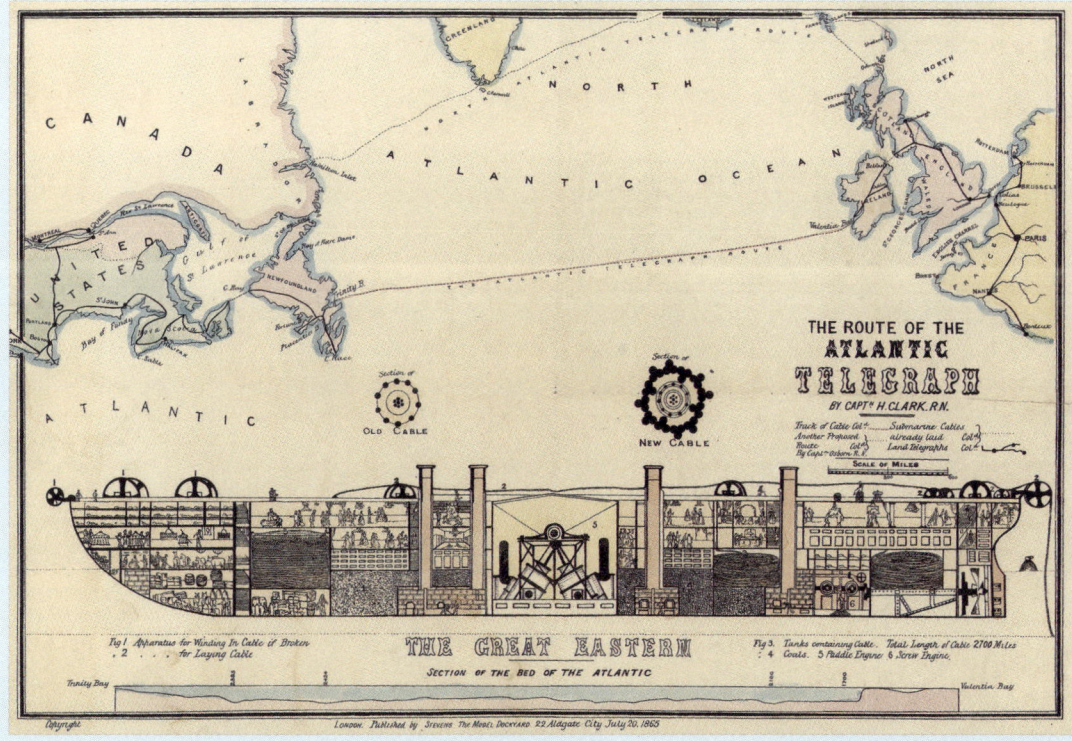

that year he sketched out a proposal for a series of wires connecting the Observatory to major railway stations that extended into the national telegraph network. His idea was approved by the South Eastern Railway and its telegraph engineer, Charles Walker, who was already involved in early cable trials across the Channel in the 1840s and who later became a useful ally in Airy's telegraphic projects.

The longitude programme began in 1852 with the installation of the Shepherd Motor Clock (49) and the successful completion of the cross-Channel submarine cable that enabled Greenwich and Paris astronomers to share time signals for the first time. The observer using the Airy Transit Circle (47) could now record his observations as a visual trace with the tap of an electrical contact, which transmitted the signal to a chronograph located in the eastern summerhouse, next to Flamsteed House. Additional telegraphic connections enabled Greenwich astronomers to share their timings with observatories in Brussels in Belgium, Harvard in the United States and the telegraph station at Valentia in Ireland, the westernmost point in Europe, as a point of reference. These experiments became an important part of the planning process in selecting and accurately measuring locations for the 1874 transit of Venus (see *Transit of Venus expeditions*). Signals from Greenwich entered the submarine cable network at Porthcurno in Cornwall and were either sent across the Atlantic via Ireland or else to Gibraltar for onward transmission across the Mediterranean. The time signals were also used to monitor the dropping of the time ball at Gibraltar.

### A continuing legacy

As British submarine networks expanded and encircled the globe, Greenwich became an integral part of this technological marvel, both in the production of the cable at Enderby Wharf and its needle-twitching pulses (56) of time emitted from the Royal Observatory. Despite a century of mergers and company changes, cable equipment production in Greenwich still continues today with Alcatel Submarine Networks providing global communications via a fleet of seven cable ships and over 800,000km (497,000 miles) of undersea cables that sustain our growing appetite for internet access.

## PRESENTATION BOX OF SUBMARINE CABLE SAMPLES

As astronomers at Greenwich and elsewhere prepared for the forthcoming transit of Venus in 1874 (64, 65, see T*ransit of Venus expeditions*), it became clear that having precise coordinates of the best observing locations across Asia and Australasia would be essential for comparing results. Airy had already used time signals sent via telegraph networks to measure the longitudes of some European observatories but connections further afield were scarce.

Since 1864 it had been possible to send telegraphic messages from London to India via overland cables but the geopolitics of routing British cables through foreign lands led to calls for a direct submarine link. In 1869 businessman John Pender raised the necessary capital to create the British Indian Telegraph Company, and cable production soon began at Enderby Wharf in Greenwich. Impressive presentation boxes such as this one, extending 18×13in. (46×33cm), were used either to persuade investors to commit or else presented as gifts after the cable's completion. Each cable section consisted of multiple strands of copper wire sandwiched between layers of waterproof gutta percha and tar, covered in an abrasive layer of hemp soaked in silica that was intended to deter hungry teredo worms. The thickest sections were used for the shore-end connections, while the thinner cables were used out at sea. The cables were loaded onto five ships, including Brunel's colossal *Great Eastern*, that headed out to Bombay (Mumbai) in late 1869. After two months of cable-laying, the connection to Suez was secured and the full connection to Britain via Egypt, Malta and Portugal was eventually finished in June 1870.

Four years later, the cables proved their worth with a series of time signal experiments that enabled astronomers to successfully measure the longitudes of Alexandria, Suez and Aden, in Egypt and Yemen, in preparation for the transit of Venus. While the astronomical event may only have lasted a few hours, the resulting telegraph cables to Bombay and beyond helped secure the Royal Observatory's global influence among its peers for many years afterwards.

Submarine cable samples, British Indian Submarine Telegraph Company, 1870 (AAB0152)

## AIRY'S SÈVRES VASE

By the 1870s, Airy had served as Astronomer Royal for nearly 40 years and had become a key figure within the Victorian scientific establishment, often enlisted to advise on subjects beyond astronomy such as railways, lighthouses and coinage. He was appointed as President of the Royal Society for the years 1872–73 and returned to his ongoing interest in the standardisation of weights and measures. Although the Observatory was home to a set of public standards of imperial lengths (52), Airy was an advocate for using metric standards in both science and – somewhat prophetically – Britain's coinage. Originally devised by French scholars in 1795, the metric system was legally adopted in Britain in 1864 but was rarely used outside scientific contexts. In 1872 Airy travelled to Sèvres, near Paris, to represent Britain at the International Commission of the Metre, where he was presented with this personalised gift. In a classic baluster form, the imposing 38in. (96cm) tall vase has the dedication set within a laurel-and-oak-leaf crown on the characteristic royal blue (*bleu de roi*) background.

It was an impressive gesture but sadly Airy's efforts to support the metric system were dashed when the Government refused to pay for his attendance at the signing of the treaty in Paris on 20 May 1875 which would establish an international laboratory for defining standards, the Bureau international des poids et mesures (BIPM). This lack of British compliance became a bargaining chip at successive conferences: if the French must accept Greenwich as the home of longitude, surely the British should accept Paris as the home of weights and measures? As a last-minute concession, the UK eventually signed the Metre Convention and became a member of the BIPM on 17 September 1884, just weeks before the International Meridian Conference (69).

The vase was initially installed near the entrance to the Observatory's Octagon Room (2) and was later displayed at Herstmonceux (see *Epilogue*) before being transferred to the National Maritime Museum in 1998. Eagle-eyed viewers may note the unique pattern of cracked lines, which is testament to the painstaking work by conservators who glued the vase back together after it shattered into 200 pieces during the Great Storm of 1987.

Sèvres porcelain vase presented to Airy, unknown maker, 1872
(ZBA1719)

## IMPERIAL ORDER OF THE ROSE

On 25 May 1871, the Emperor of Brazil, Dom Pedro II, and his wife Teresa Cristina embarked on a tour of Europe as they joined relatives in Portugal to mourn the death of their 23-year-old daughter Leopoldina. Despite the poignant circumstances, the Emperor decided to use this voyage as an opportunity to indulge in his love of science by visiting key sites across Europe. With his own private observatory and laboratory for making daguerreotypes (an early type of photograph) back in Rio de Janeiro, the Emperor was keen to make a pilgrimage to Greenwich, finally arriving on 17 July.

Reporting to the Board of Visitors a year later, Airy recalled how his distinguished guest 'minutely examined every part' of the site. At that time, Airy was busy preparing for the forthcoming transits of Venus, predicted for 1874 and 1882 (64, 65, see *Transit of Venus expeditions*). He explained to the Emperor how the first event would not be visible in Rio but promised to keep him informed about the second event. Neither man could speak the other's language well so they had to converse in a mixture of French and English.

The Emperor was impressed by his visit and awarded Airy the Grand Cross of the Imperial Order of the Rose, seen here. It was a generous and prestigious gift that clearly delighted Airy, leading him to respond: 'With pride I receive this proof of Your Majesty's recollection of your visit to the scientific institutions of Great Britain.' The Order of the Rose was originally created in celebration of the marriage of the Emperor's parents, Dom Pedro I and Amélie of Leuchtenberg, in 1829. The medallion features the gold entwined initials 'P & A' surrounded by the motto 'AMOR E FIDELIDADE' ['Love and Fidelity'] and a circle of roses, said to be Amélie's favourite flower. This award was one of many presented to Airy over the course of his career but seemingly occupied a special place in his heart, inspiring him to welcome guests from the Brazilian Navy in the 1880s, even after retirement.

Neck badge and collar star, unknown maker, made before 1872
(MED2116)

## 64
## DALLMEYER PHOTOHELIOGRAPH

This small but unusual telescope bears witness to the beginning of a new era of observation at Greenwich. In 1857, chemist and amateur astronomer Warren de la Rue created a pioneering type of telescope, known as a photoheliograph, which enabled him to take daily images of the Sun and its changing pattern of sunspots. At the same time, Airy was devising observing plans for the next transit of Venus (see *Transit of Venus expeditions*) in 1874. As the first such event since the invention of photography, it was an ideal opportunity to take advantage of the new technology. Five photoheliographs were ordered from the London-based optical company of Dallmeyer and, after initial tests at Greenwich, were shipped to observing stations in Egypt, New Zealand, Hawaii and the Indian Ocean islands of Rodrigues and Kerguelen.

This specific instrument, Dallmeyer no.2, is thought to have been sent to New Zealand and then used at Greenwich from 1910 to 1949. Sunlight was directed through the 4in. (10cm) aperture lens onto a magnifying screen that enlarged the 0.6in. (1.5cm) image by about seven times. With such intense light and heat, the operator used a spring-loaded slit to minimise the exposure time to just a fraction of a second. He then extracted and replaced the photographic plate with a fresh one. The enlarging lens and mahogany camera box seen on this instrument today are later additions from 1909.

Despite years of planning, the transit of Venus expedition of 1874 was deemed a failure with poor quality photographs and inconclusive results. It was a disappointing outcome after years of planning but the Observatory still reaped the benefits by continuing to use the Dallmeyer photoheliographs for decades afterwards. Recruited a year before the transit of Venus, E. Walter Maunder (67, 72, 80, 81, see *Splitting starlight with spectroscopy*) was responsible for the new Heliographic Department and spent the next 40 years using various Dallmeyer photoheliographs to take daily photographs of sunspot patterns. He relied on images taken by astronomers using similar instruments in India, South Africa and Mauritius to complete the record when the skies were cloudy at Greenwich. This invaluable data set is still used by scientists today.

A photoheliograph in one of the temporary huts installed at Greenwich to help train observers before the transit of Venus expedition, unknown maker, about 1874
(AST0968, B1636-17A)

Photoheliograph, J.H. Dallmeyer, 1873 (AST0968)

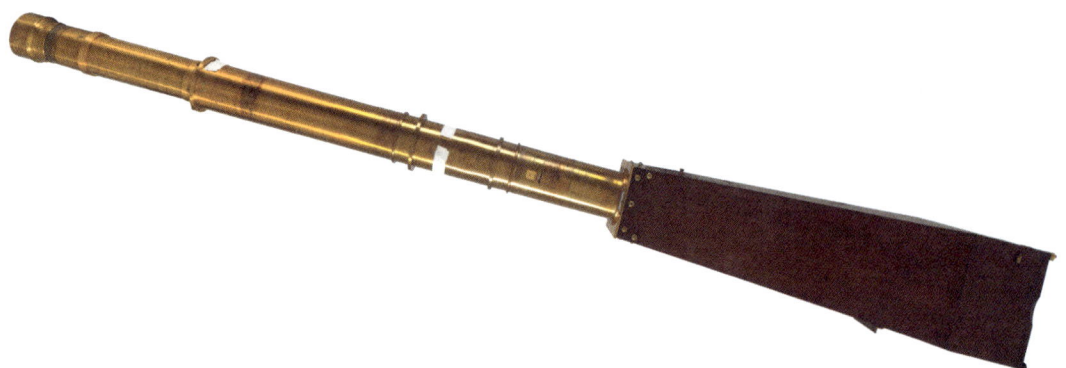

170 A HISTORY IN OBJECTS

## 65

## TEST PLATE FOR THE JANSSEN PHOTOGRAPHIC REVOLVER

Although conventional photography made little difference to the results of the 1874 transit of Venus expeditions (see *Transit of Venus expeditions*), the event did inspire the birth of the movie camera. In February 1873, French astronomer Jules Janssen developed an idea for a photographic revolver by which a silvered copper disc and adjustable shutter connected to a clockwork drive could be used to take a sequence of images. He realised this approach could help record the exact moment at which Venus appeared to enter and exit the Sun's disc, which was vital to measuring the Sun–Earth distance but notoriously difficult to record. With approval for his design from the French government, Janssen sailed to an observing site in Japan, narrowly escaping two typhoons en route. On arrival at Nagasaki, he chose a hilltop location near the harbour, Kompira-Yama, and persuaded 500 porters to carry over 250 boxes of instruments to the summit.

But the Frenchman was not the only one using the photographic revolver: by April 1873, his device had already caught the attention of Airy who commissioned the manufacturer of the transit photoheliographs (64) to create something similar for the British expeditions. As the countdown to the transit continued, Airy installed a model on the roof of Flamsteed House to help his teams practice their observations. It consisted of a metal disc (Venus) slowly propelled by clockwork – at less than 0.001in. (0.0025cm) per second – across a mock solar disc illuminated by sunlight. This circular glass photopositive and 44 identical images is believed to have been one of the practice plates used with this model.

For the transit itself, more than nine Janssen-type photographic revolvers were used by various French and British observing parties but the plates produced were only partially successful, either due to cloudy conditions or technical issues. With so few results worthy of analysis, the revolver was discontinued for the second set of expeditions in 1882. Nonetheless, its pioneering principle of sequential photography lived on in the emergence of movie cameras, for which Janssen is rightfully credited.

Test plate for the Janssen photographic revolver, 1874 (AST1081)

Illustration of Janssen's revolver in use, *La Nature*, 1875 CNUM - Conservatoire numérique des Arts et Métier

# SUNSHINE RECORDER

As Walter Maunder settled into the new routine of taking daily photographs of the Sun (64), his colleagues William Ellis and William Nash were busy measuring the duration of sunlight with a new gadget that meteorologists continue to use today. In 1876, Scottish aristocrat John F. Campbell presented the Observatory with a version of the sunshine recorder that he had devised over 20 years previously, possibly a continuation of his childhood fascination with the burning properties of glass. The instrument consisted of a brass bowl with a strip of blackened cardboard that was replaced each sunset and positioned underneath a 4in. (10cm) diameter glass sphere. Once the bowl was aligned with the meridian, the sphere focused the Sun's rays onto the card, leaving a scorched trace that represented the hours of bright sunlight, hence the motto around the rim: HORAS NON NUMERO NISI SERENAS ['I only count the serene hours'].

The instrument was installed on 7 May 1876 and remained in use until 1886 when it was replaced with an improved version created by George Gabriel Stokes, in which the strip was clamped securely within a metal frame, making it easier to record and measure the results more accurately. The new sunshine recorder was positioned next to the rain gauge (37) on a rooftop platform on the Magnet House (46), but was moved to the roof above the Octagon Room (2) in February 1896 to avoid disturbance from the construction of the nearby New Physical Observatory (78).

On average, the sunshine recorder traced around 1,200 hours of bright sunshine each year, about 27 per cent of the possible 4,454 hours in which the Sun was visible above the horizon. At the height of summer, the traces began at 5 a.m. and continued until 8 p.m., whereas in the winter the traces only extended for three hours around noon. By the 1930s, however, the sunshine results had started to decline and the tenth Astronomer Royal, Harold Spencer Jones, used the data as evidence of 'atmospheric impurity' to argue in favour of relocating the Observatory to a better observing site.

Sunshine recorder bowl, J.F. Campbell, 1876 and spherical lens, Chance Brothers Ltd, 1950
(AST0770, AST0769)

IN FOCUS

# Splitting starlight with spectroscopy

While other scientists can directly collect and analyse their objects of study here on Earth, astronomers are entirely dependent on light – both the visible and invisible parts of the electromagnetic spectrum – as their only source of information about the stars. In the 1860s, spectroscopy, a technique originally developed by chemists in laboratories, opened up a completely new realm of scientific enquiry for astronomers. They could now use starlight to determine a star's mass, age, chemical composition and motion relative to Earth. Although the spectroscopic work at the Royal Observatory was not particularly innovative compared to similar institutions, the surviving notebooks reveal the true story of how astronomers at Greenwich battled with and finally embraced this revolutionary new technology that transformed astronomy into modern astrophysics.

**Investigating sunshine**
Spectroscopy was inspired by observations of our closest star, the Sun. In 1672, Isaac Newton directed sunlight shining through a hole in a shuttered window through a prism to demonstrate how light could be split into a continuous spectrum across seven coloured sections, and then reassembled back into white light with a lens. Over 130 years later, the chemist William Hyde Wollaston repeated Newton's experiment with a smaller, slit-shaped aperture and found that the colours were separated by dark lines. As natural philosophers tried to understand these optical phenomena, craftsmen working with lenses added their own insights, notably the Bavarian optician Joseph Fraunhofer. In a bid to minimise colour dispersion in his products, Fraunhofer used a

pure colour (monochromatic) sodium lamp to provide a comparison. By viewing the element's distinctive bright line in the yellow-orange part of the spectrum through different lenses, he could assess the quality of his products: the stronger the line, the better the lens. He created the spectrum by viewing the emitted light through a prism and telescope from a theodolite, effectively creating the first 'spectroscope' (67). When he tried this apparatus with sunlight, he was astonished to see 574 lines, which were later named 'Fraunhofer lines' in his honour. The optician named the darkest lines 'A' and found that the 'D' line coincided with the bright line seen in the sodium spectrum. He later turned his technique to the skies and identified similar spectral lines for the Moon and planets, confirming that their light was indeed reflected sunlight, although he found that starlight was much more varied. Fraunhofer ceased his own astronomical investigations and focused on his business, which became renowned across Europe for its high-quality telescopes.

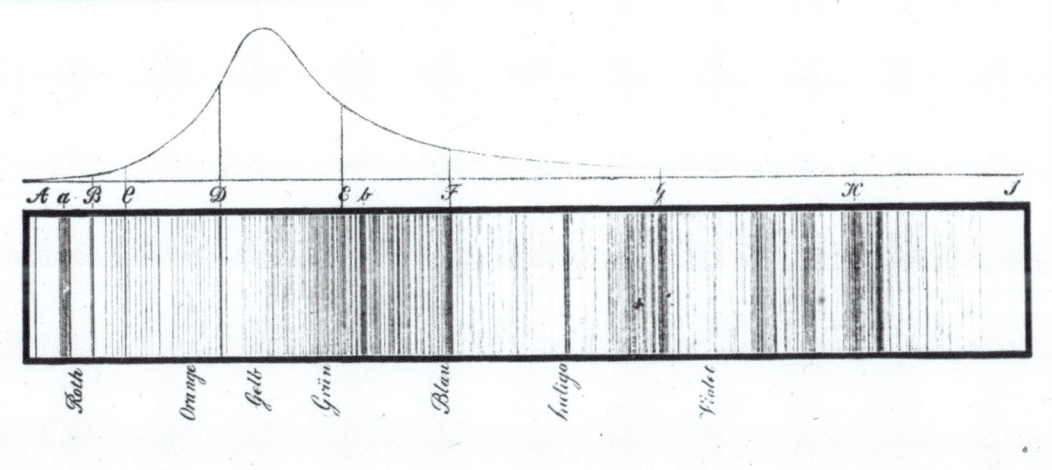

← Lantern slide showing a chemist analysing a burning sample (right) with a laboratory spectroscope. Light from a burning candle (centre) was directed through a tiny slit for comparison, unknown maker, about 1880 (ZBA4535.25)

↑ Lantern slide showing the brightness curve above Fraunhofer's spectral lines derived from sunlight, from red (left) to violet (right), unknown maker, about 1880 (ZBA4535.28)

**New insights from the laboratory**

After Fraunhofer's death in 1826, spectroscopy dwindled to a niche topic that was only explored by curious and wealthy enthusiasts. The next phase of the subject's development came from the industrial heartlands surrounding the River Rhine where chemists sought to improve manufacturing processes and create new products. In 1852, Robert Bunsen – of laboratory burner fame – was appointed director of a new chemical institute at the University of Heidelberg. Working in collaboration with Gustav Kirchhoff, Bunsen started to analyse burning chemical samples with his own version of a spectroscope, in which emitted light was directed through a reversed telescope and prism, with the resulting spectrum observed through a second telescope. By 1859 the two chemists had deduced that bright emission lines in spectra were generated by hot, burning gases whereas dark lines indicated the absorption of light by cooler gases. As an increasing number of elements could be identified by their own unique set of spectral lines, Bunsen and Kirchhoff confidently announced to their peers in 1860: 'It is evident that the same mode of analysis must be applicable to the atmospheres of the Sun and brighter stars.' It was now time for astronomers to extrapolate this technique from the lab bench to the telescope.

**Analysing starlight**

In the aftermath of Kirchhoff and Bunsen's announcement, astronomers around the world took up the challenge to classify stars according to their chemical spectral signature. One of the first proponents was Father Angelo Secchi of the Roman College Observatory, who systematically catalogued over 4,000 stars into four main spectral types. These were refined further by another project organised by Edward Pickering at Harvard College Observatory in the 1890s. Pickering relied on a team of women astronomical computers, some of whom

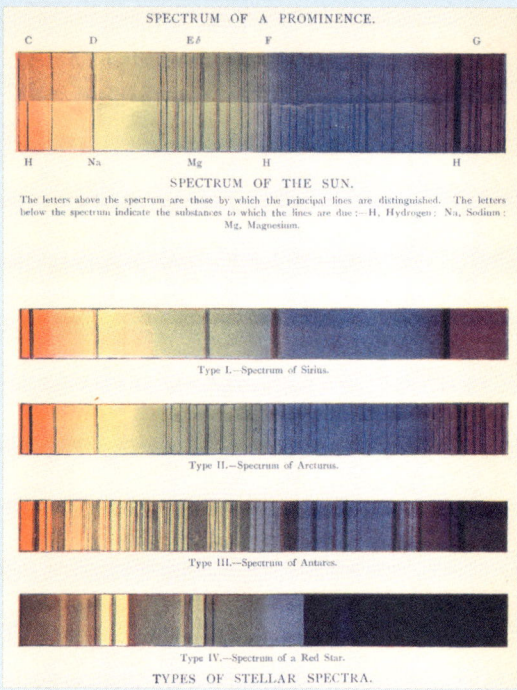

reputedly became proficient at cataloguing a single stellar spectrum within three seconds. The resulting list of 10,000 stars by 13 spectral types laid the foundation for the classification system used by astronomers today and transformed our understanding of stellar composition and evolution.

In addition, astronomers realised that spectroscopy could help identify the nature and motion of the stars. For example, in 1864, wealthy English amateur astronomer William Huggins obtained a single bright emission line of a gas cloud visible in the constellation of Draco – known today as the Cat's Eye Nebula – that confirmed it was composed of luminous gas, rather than an aggregation of stars. Eight years later, Huggins used the spectra of Sirius and other stars to confirm Christian Doppler's 1842 theory that spectral lines are shifted towards the red or blue end of the spectrum if a light source is moving away from or towards us, respectively.

The application of spectroscopy developed in parallel with new technologies. In 1868 Swedish astronomer Anders Ångström created a new solar spectrum profile with over 1,000 lines by exchanging the prism for a diffraction grating, a plate covered in very fine grooves or slits that dispersed the light more widely, and made the spectral lines easier to identify. Just four years later, New York astronomer Henry Draper photographed the bright star Vega with a diffraction grating to capture the first photographic stellar spectrum (spectrogram). It was the start of a close association between photography and spectroscopy that continues to shape astronomy today.

### Spectroscopy at Greenwich

At first, the adoption of spectroscopy was rapid with some initial trials in 1862 using a spectroscope attached to the Observatory's largest telescope at the time, the 12.8in (33cm) Merz refractor (75). The early results were disappointing and a dedicated spectroscopic programme did not begin until the creation of the Heliographic and Spectroscopic Department in November 1873, headed by E. Walter Maunder. As preparations continued for the forthcoming transit of Venus expedition (see *Transit of Venus expeditions*), Maunder combined his daily photographs of the Sun with spectroscopic observations of sunspots (dark patches) and large filaments, extending from the Sun's atmosphere (prominences). He also worked with Chief Assistant William Christie (later Astronomer Royal) on detecting carbon dioxide in the tail of

Illustration of comparative solar and stellar spectra in *The Heavens and Their Story*, Annie and Walter Maunder, about 1908 (PBH8641)

William Huggins' Observatory, London, in *An Atlas of Representative Stellar Spectra*, Sir William and Lady Huggins, 1899 (PBG1302)

E. Walter Maunder uses a spectroscope attached to the 28in. Great Equatorial Telescope while William Bowyer takes notes, attributed to David Edney, about 1895 (REG18/000454.67)

Coggia's Comet. It was complicated and messy work that involved an assistant producing an electric spark using a smelly acidic battery and an induction coil. This spark ignited an evacuated glass tube refilled with a known gas. The observer then tried simultaneously to observe and compare the reference spectrum in the tube and the spectrum from the celestial object in the eyepiece. Despite their best efforts, Maunder and Christie were hampered by faint spectra, weak batteries, poorly focused spectral lines and windy conditions that distorted their view. The surviving notebooks reveal their frustration and dejection as they grappled with the new technology, seen most spectacularly on 29 July 1874: 'Something entirely wrong with these observations. Rejected.'

The arrival of the 28in. (71cm) Great Equatorial Telescope (75) in 1893 seemingly provided the ideal opportunity to combine spectroscopy with photography, but the initial trials were unsuccessful and the spectroscope was later reassigned to the 30in. Thompson Reflector (82). Apart from some opportunistic spectrograms taken during eclipse expeditions in the first decade of the twentieth century, spectroscopy was marginalised at Greenwich until after the First World War. In 1922, Greenwich astronomers Charles Davidson and Ernest Martin began a programme to measure the colour of 25 stars. Each star emits radiation that produces a continuous mountain-shaped curve of energy versus wavelength (Planck's 'black-body radiation'). The peak of the curve is known as the star's effective wavelength and provides a measure of colour and surface temperature.

By analysing a select number of well-known star types, Davidson and Martin hoped to create a zero (absolute) point for comparison across a bigger range of star colours, from cool, red stars to hot, blue stars. They used a spectrograph attached to the Thompson Reflector and relied on precisely calibrated lamps installed on the roof of Flamsteed House to provide a reference spectrum. A decade later, the programme continued with new spectrographs (96) on the 36in. (91cm) Yapp Telescope that enabled them to extend their research into ultraviolet wavelengths. The subsequent publication of the catalogue in 1934 was a bittersweet ending as Davidson and his colleagues conceded defeat to the worsening atmospheric conditions, concluding: 'It is not proposed to undertake further absolute observations at Greenwich.' It was time to hand over the baton to astronomers with more sensitive instruments under clearer skies.

# TWO-PRISM SPECTROSCOPE

By the time Airy created the Spectroscopic Department in the early 1870s (see *Splitting starlight with spectroscopy*), new instruments were making it easier for astronomers to apply the analytical technique to their observations. This spectroscope was used with the 15in. (38cm) refractor of the Stonyhurst College Observatory in Lancashire and was made by the London optical instrument maker Adam Hilger, who supplied similar instruments to the Royal Observatory. Once the instrument had been slotted onto the telescope eyepiece, the observer inserted the relevant prism, each with a different optical density, to select the wavelength of interest.

In November 1873, E. Walter Maunder (72, 80, 81) began work alongside Chief Assistant William Christie, who designed his own spectroscope that could be fitted with extra prisms for greater dispersion of the lines. The duo started with the spectral analysis of solar features, planetary atmospheres and the tails of comets. A year later, they expanded their programme by adopting a technique developed by amateur astronomer William Huggins, who also served on the Observatory's Board of Visitors. Nearly 20 years previously, Huggins had shown that shifts in stellar spectra corresponded with a star's motion through space, as seen via our line of sight (radial velocity, see *Glossary*). Similar to the rise and fall in pitch of a siren on a moving emergency vehicle (called the Doppler effect), stars moving towards us produce spectral lines shifted towards the blue part of the spectrum, while those moving away are shifted towards the red. In 1885, Maunder and Christie detected an apparent blueshift in the motion of the star Sirius, which implied it was moving towards Earth and completely contradicted prevailing opinion. When the spurious result was eventually dismissed, Huggins and his astronomer wife Margaret expressed their support for Maunder's observational expertise and firmly placed the blame on Christie's spectroscope.

Despite the Greenwich astronomers' enthusiasm to engage with astrophysics, early spectroscopic observations were impeded by hazy skies, uncalibrated instruments and occasional fires caused by stray electric sparks. As Maunder wryly commented, 'The principle is clear enough. The actual working out of the observation was one of very great difficulty.'

➔
Two-prism astronomical spectroscope with extra prisms, A. Hilger, 1874–97 (AST1078)

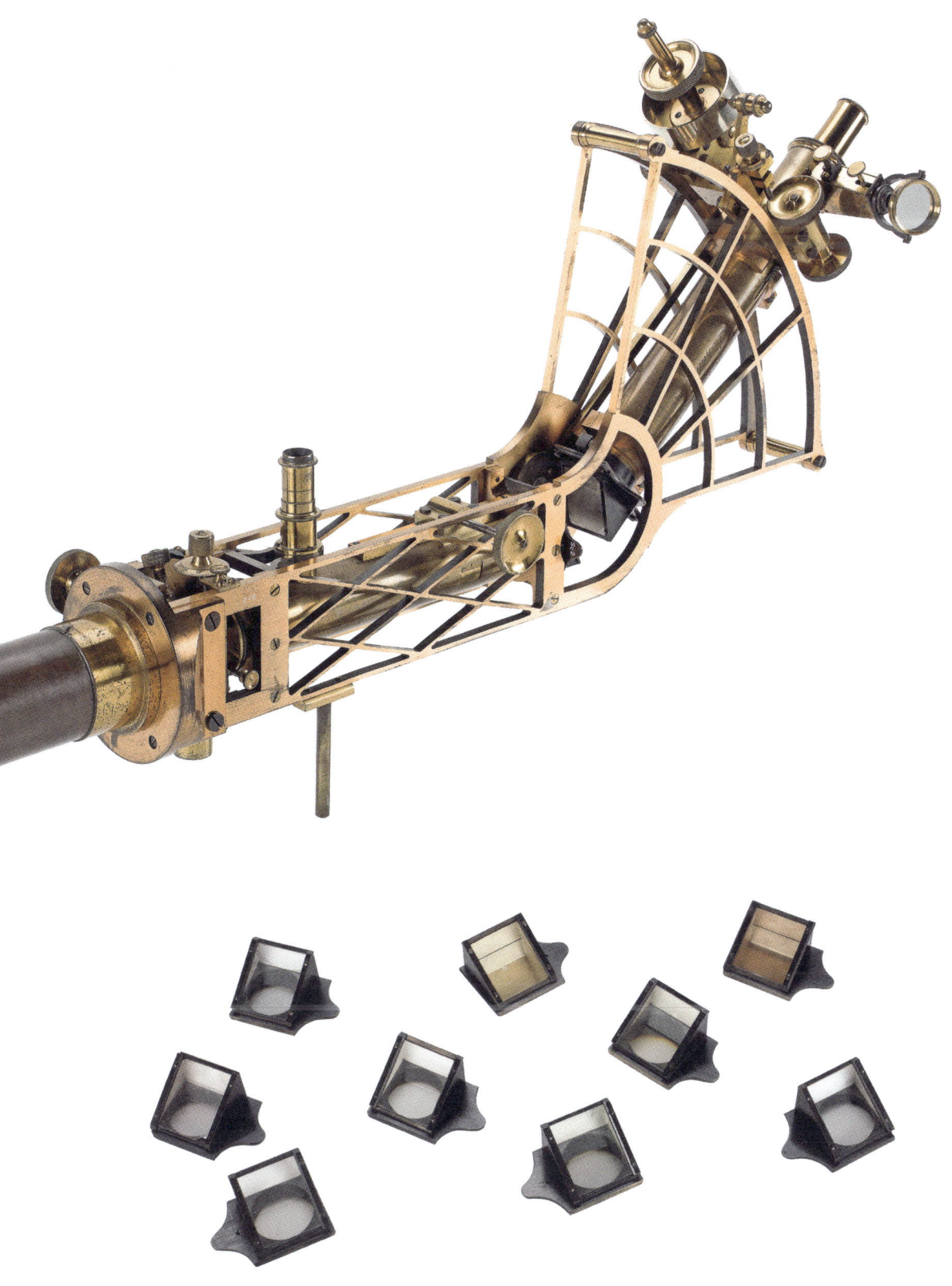

The Merz Equatorial Telescope shown as *The Great Equatorial Telescope in the Dome, [Greenwich Observatory]*, unknown artist, based on a drawing by L. Cornellisen, engraved by Vincent Brooks, Day & Son Ltd, about 1870
(AST0019)

# KEW-PATTERN UNIFILAR MAGNETOMETER

By the end of Airy's tenure as Astronomer Royal, Greenwich was well-established within a global network of magnetic observatories. The 'Mag and Met' department was now headed by William Ellis, who had joined the Observatory in 1841 as a computer at the age of 13. He was promoted after the departure of James Glaisher in 1874 and was aided by William Nash alongside a group of four computers. Despite the use of photographic paper on revolving drums to record automatically the magnetic and meteorological data, there was still a busy schedule of visual observations every four hours – except on Sundays – to provide a baseline reference.

Magnetometers such as this example were installed in the arms of the cruciform Magnet House (46). Designed to detect vertical and horizontal variations in the Earth's magnetic field, the instrument was mounted on a tripod and adjusted using the spirit level and foot screws. The glass tube housed silk threads that supported two cylindrical magnets suspended within a wooden box to protect them from draughts. With a lamp and mirror for illumination, the observer could use the sighting telescopes to measure the magnet's variable movement (time of vibration) against the finely calibrated scales. He could also use the adjustable mirror to check the instrument's alignment with the local meridian, as measured by the reflection of the north star (Polaris) seen through a roof hatch.

This type of magnetometer was known as 'Kew-Pattern' because it was made according to the standards devised by Kew Observatory. The west London institution was originally built for George III to observe the transit of Venus in 1769 (see *Transit of Venus expeditions*) and to house his collection of scientific instruments. It became renowned as a centre of expertise for both collecting magnetic and meteorological data and for the testing and certification of watches and scientific instruments, for which manufacturers gladly paid testing fees to secure the prestigious 'Kew certificate'. The venue was administered by various scientific organisations until it became part of the National Physical Laboratory in 1900.

◉
Kew-Pattern unifilar magnetometer, Elliott Bros, 1870–90 (AST0761)

# Sir William Henry Mahoney Christie

## 1845–1922

Born locally in Woolwich, William Christie started work at the Observatory as Chief Assistant on 9 September 1870. The 24-year-old had performed well in the Mathematical Tripos at Cambridge, and Airy – one of the examiners – duly offered him the role. As Airy's deputy, Christie's main task was to check the previous night's observations, along with certifying new observers and making a few monthly observations with the Airy Transit Circle. Unusual events such as the appearance of comets and exceptionally large sunspots gave him the opportunity to pursue his interests in spectroscopy.

When Airy retired in August 1881, his post was originally offered to the Cambridge mathematician John Couch Adams, famous for his involvement in the discovery of Neptune. However, he declined, and Christie accepted instead. Within a few months, the new Astronomer Royal's life changed again with his marriage to Violette Mary Hickman, who tragically died shortly after giving birth to their second son in 1888.

Having worked at Greenwich for over a decade, Christie understood the limitations of the site and set upon an ambitious programme of improvements. In 1885 he successfully persuaded the Admiralty to fund a much larger, 28in. (71.2cm) diameter refracting telescope, a feat that he memorialised within his portrait. He also instigated the building of the New Altazimuth Pavilion and attracted offers of large telescopes from wealthy donors. Similarly, the Carte du Ciel photographic project in 1887 was an ideal opportunity to boost the Observatory's global profile. He then turned his attention to the site's infrastructure with the construction of the New Physical Observatory, followed by the relocation of the magnetic instruments to an enclosure about 300ft (400m) away in Greenwich Park, which later became the site of the Yapp Telescope in the 1930s.

But, despite the significant investment, time was running out for observations at Greenwich as London's infrastructure encroached on the site, most notably with the construction of Greenwich Power Station directly on the Prime Meridian in 1906. Overstretched and increasingly unwell, Christie retired in 1910 and moved to Surrey. In January 1922, he suffered a gastric haemorrhage while on a cruise to Morocco and was buried at sea near Gibraltar.

*Sir William Henry Mahoney Christie*, G.P. Jacomb-Hood, 1911
(PAH5622)

THE INTERNATIONAL MERIDIAN CONFERENCE,

WASHINGTON, 1884.

Baron von ALVENSLEBEN, *Minister of Germany.* M. de J. GALVAN, *Minister of San Domingo.* A. LEFAIVRE, *Consul-General of France.* Baron von SCHÆFFER, *Minister of Austria.*

Mr. HINCKELDEYN, *Germany.* J. VALERA, *Minister of Spain.* A. M. SOTELDO, *Venezuela.* Lt.-Gen. STRACHEY, Secretary, *Great Britain.* M. JANSSEN, Secretary, *France.* Admiral C. R. P. RODGERS, *President.* Dr. L. CRULS, Secretary, *Brazil.* G. H. de WECKHERLIN, *Minister of Netherlands.*

Col. E. FREY, *Minister of Switzerland.* L. M. RUTHERFURD, *United States.* Comdr. FRANKLIN, *Colombia.* Sir F. J. O. EVANS, *Great Britain.* Prof. J. C. ADAMS, *Great Britain.* W. F. PEDDRICK, *Reporting Secretary U. S.* RUSTEM EFFENDI, *Turkey.* Count FORESTA, *Italy.*

J. KIKUCHI, *Japan.* L. FERNANDEZ, *Mexico.*

CLEVELAND ABBE, *United States.* J. de KOLOGRIVOFF, *Russia.* C. de STRUVE, *Russian Minister.* Maj. Gen. STEBNITZKI, *Russia.* Comdr. W. T. SAMPSON, *United States.* A. ANGUIANO, *Mexico.* A. A. ADEE, *Ass't Sec'y State of U. S.*

JUAN PASTORIN, *Spain.*

F. V. GORMAS, *Chili.* A. S. TUPPER, *Chili.* Count LEWENHAUPT, *Minister of Sweden-Norway.* M. A. BATRES, *Minister of Salvador.* M. M. BOCK, *Guatemala.* L. AHOLO, *Hawaii.* Capt. J. STEWART, *Consul-General of Paraguay.* F. R. MARCEAU, *French Stenographer.*

J. F. ECHEVERRIA, *Costa Rica.* W. D. ALEXANDER, *Hawaii.*

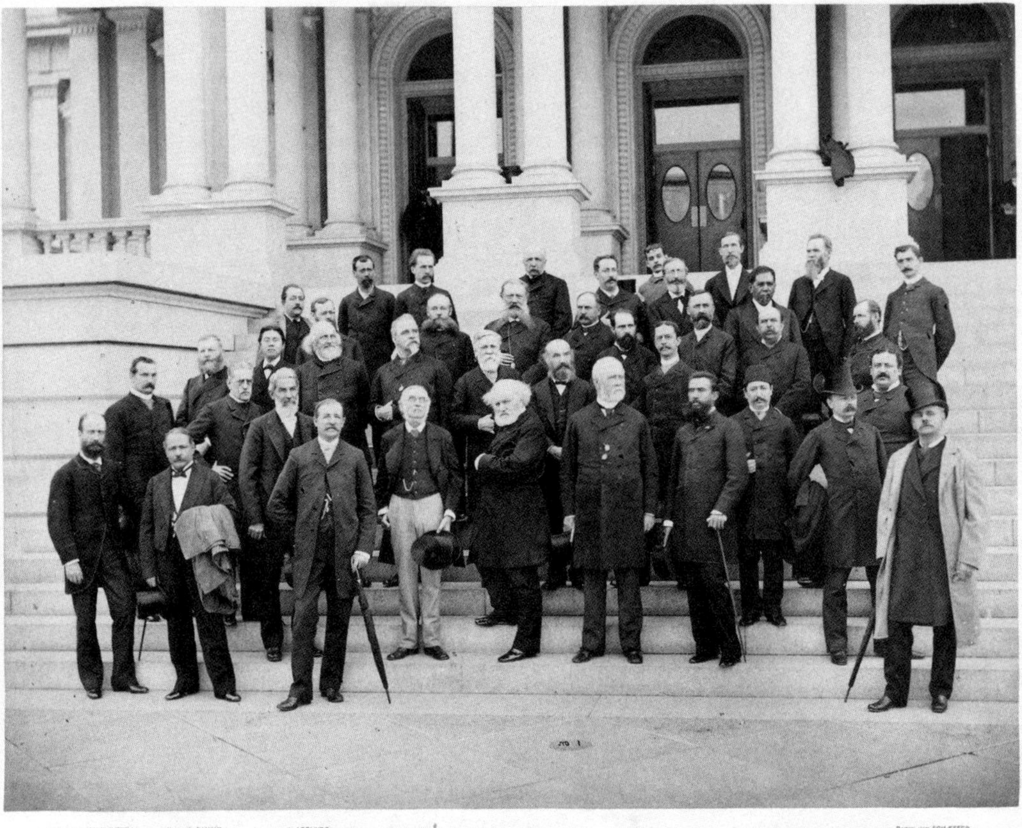

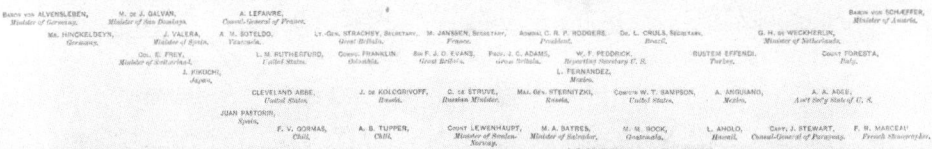

## THE 1884 INTERNATIONAL MERIDIAN CONFERENCE

It had been an exhausting fortnight of meetings and debate but finally, on 13 October 1884, delegates at the US State Department agreed that the meridian defined by the Airy Transit Circle (47) at Greenwich should be used as the Prime Meridian of the World (0° longitude). It was the culmination of several years' wrangling by telegraph operators and US railroad companies to coordinate global trade, travel and communications.

By the early 1880s, there were at least 11 different prime meridians defined by observatories in Europe and the Americas. Initial discussions among scientists at a conference in Rome in October 1883 had confirmed various nations' desire for unity in time and longitude but ended with no clear outcome. A year later, the US Government invited 41 delegates from 25 countries to Washington D.C. 'for the purpose of fixing upon a meridian proper to be employed as a common zero of longitude and standard of time-reckoning throughout the globe'. The delegates were a mixture of scientists, diplomats, engineers and military personnel; the Astronomer Royal did not attend.

At first, the French delegates proposed a neutral meridian through the Pacific, but the others argued that the prime meridian should be determined by a national observatory with the best instruments to guarantee its accurate definition and longevity. Eventually, four national observatories came under consideration, but it was the compelling statistics from Sandford Fleming, a Scottish-Canadian railway engineer, that swung the debate in favour of Greenwich. According to Fleming, the majority of shipping companies (72 per cent by tonnage) were already using the Greenwich meridian, making its universal adoption the most convenient and economic choice. With over a century of use of the *Nautical Almanac* (19) and several decades of Royal Navy surveying expeditions using Observatory-rated chronometers (51), the Greenwich meridian was already firmly embedded within charts and navigational practice.

The resolution was passed by 21 affirmative votes, with one objection from San Domingo and two abstentions from France and Brazil. Despite only being a recommendation with no legal bearing, it was adopted by most countries within 20 years. The conference itself continued for another two weeks as the discussions switched focus to the creation of a universal day (70).

Delegates at the International Meridian Conference, Washington D.C., unknown photographer, 1884 (D9643)

## 70
## WORLD TIME CONVERTER

Having resolved to adopt the Greenwich meridian as Prime Meridian of the World, delegates at the 1884 International Meridian Conference (69) turned their attention to time. They were asked to consider Resolution IV that proposed 'the adoption of a universal day'. Just a year earlier, US and Canadian railroad companies had adopted four standardised time zones – as initially devised by Professor Charles Dowd in 1874 – and they advocated for a similar approach for global time. The resolution was unanimously passed and led to Resolution V, which stated that the universal day should start at midnight at the Prime Meridian, with Universal Time defined by Greenwich Mean Time. The Greenwich meridian was now officially the beginning of both time and longitude, although it would take several decades for nations to adopt these recommendations and for the now familiar system of global time zones to emerge.

This colourful world time calculator from the period of the conference demonstrates how people tried to adapt to these changes. Invented by C. Pascal, this gadget is based on the Paris meridian with places behind Paris time shown in red and those ahead in yellow. With rotating scales subdivided into intervals of 0.5° longitude, the user could calculate the difference in longitude and local time between locations, knowing that the Earth rotates 1° every four minutes. Aligning the blue hour scale to the required locations could also be a quick reference, such as 8 p.m. in Athens is equivalent to 4 a.m. in Brisbane.

For Pascal and his compatriots, Paris Mean Time as defined by the city's observatory remained the basis of French standard time until March 1911 when the country switched to GMT to align with the transmission of international radio signals. French lawmakers preserved their capital's astronomical heritage by describing the new time standard as 'Paris Mean Time diminished by 9 minutes 21 seconds' to denote the longitude difference between Greenwich and Paris. Like several other countries then under Nazi occupation, France advanced its clocks in 1940 to align with Berlin time and remains an hour ahead of GMT today.

➲ (and overleaf)
World time converter
(and rear view), C. Pascal,
1875–1900 (AST0700)

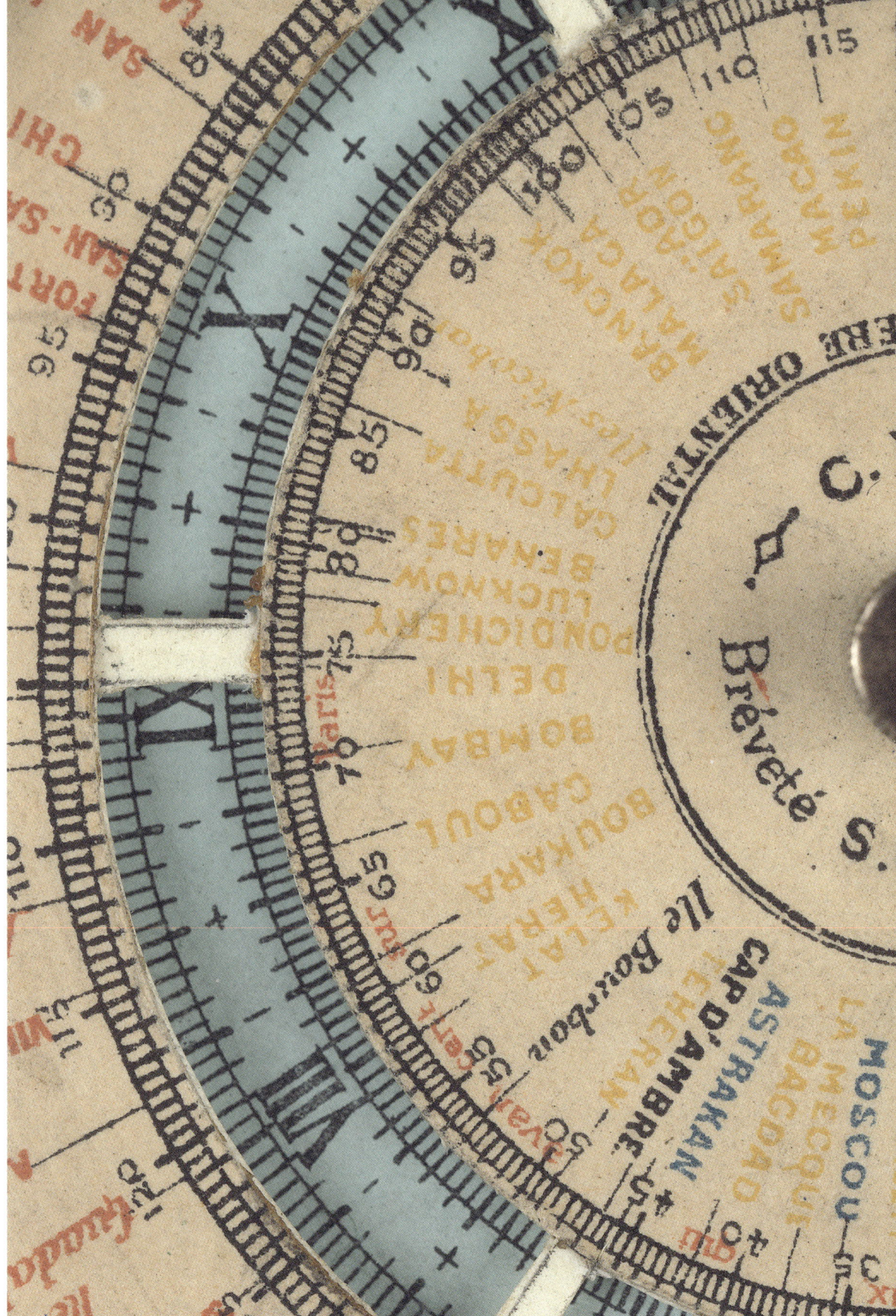

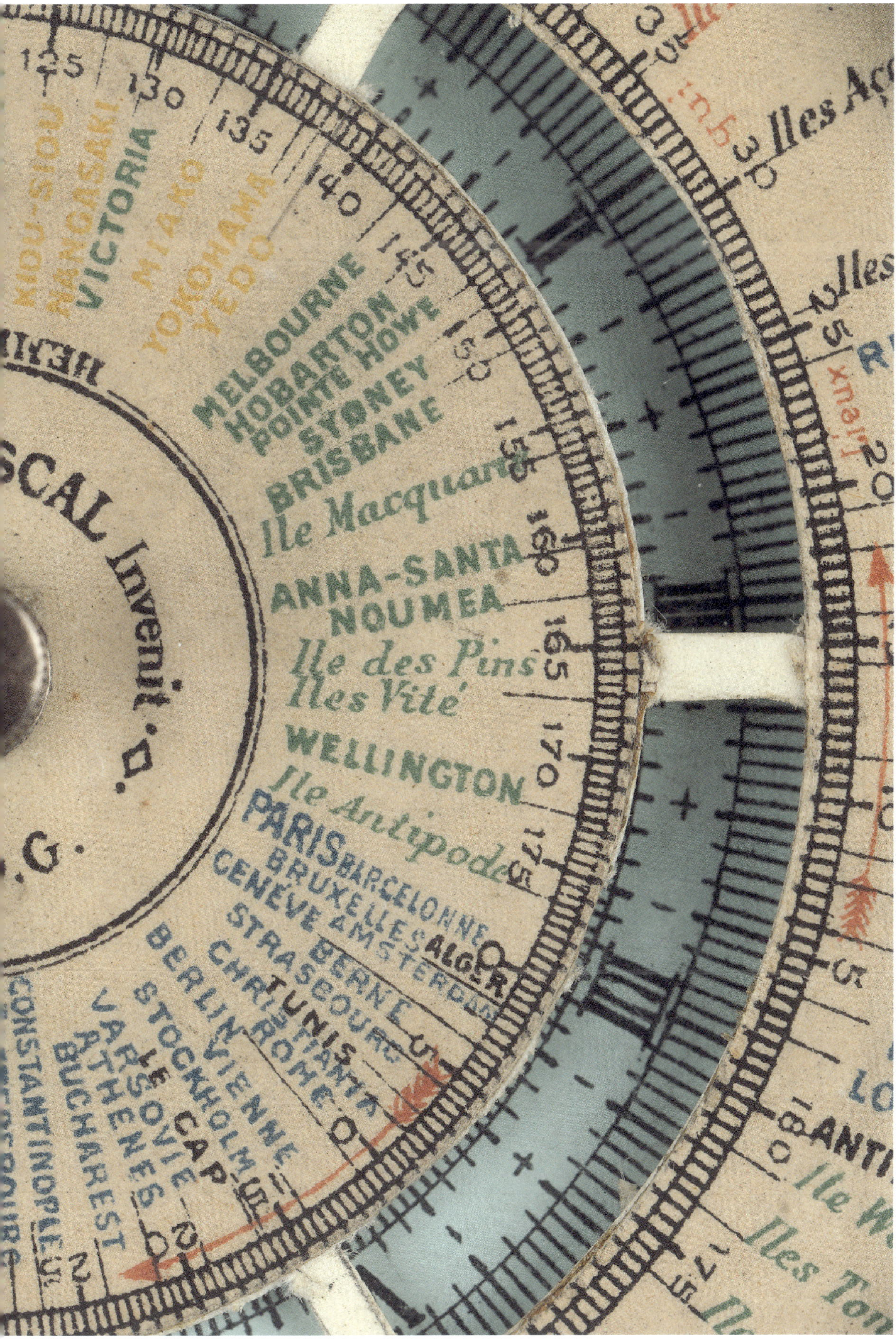

KIOU-SIOU
NANGASAKI
VICTORIA
MIAKO
YOKOHAMA
YEDO

MELBOURNE
HOBARTON
POINTE HOWE
SYDNEY
BRISBANE
Ile Macquarie

ANNA-SANTA
NOUMEA
Ile des Pins
Iles Viti
WELLINGTON
Ile Antipode

PARIS
BARCELONNE
BRUXELLES
GENÈVE AMSTERDAM
ALGER
STRASBOURG
BERNE
TUNIS
CHRISTIANIA
ROME
VIENNE
BERLIN
STOCKHOLM
LE CAP
VARSOVIE
ATHÈNES
BUCHAREST
CONSTANTINOPLE

PASCAL Invenit.

Iles Aç...
Iles ...
Rx...
ANTI...
Ile W...
Iles Ton...

Royal Observatory, Greenwich. PLATE I.

ASTROGRAPHIC EQUATORIAL.

# 71
## ASTROGRAPHIC TELESCOPE

Unlike other comets which were associated with disaster, the Great Comet of 1882 was a fortuitous event that heralded a new age of collaborative astronomy. Astronomer David Gill at the Cape Observatory photographed the comet's profile and was struck by the multitude of faint stars pictured in the background. The image was circulated and eventually reached the Paris Observatory where it caught the attention of its Director, Ernest Mouchez. Keen to capitalise on the potential of photography, Mouchez enlisted the brothers Paul and Prosper Henry to devise the Carte du Ciel, an ambitious project to create a photographic atlas and catalogue of several million stars within 10–20 years.

Realising that it would be an impossible task for just one single observatory, Mouchez invited 56 delegates from 19 nations – one of whom was the Astronomer Royal Christie – to meet in Paris in April 1887. Eventually, 18 observatories agreed to participate, each one allocated a specific part of the sky and required to purchase a 13in. (33cm) diameter astrographic refractor, suitable for 6.3×6.3in. (16×16cm) glass plates for consistency. Christie ordered one from Howard Grubb of Dublin and the new instrument was installed in May 1890, mounted with a parallel set of guiding telescopes. A few months later, Christie hired four 'lady computers' as cheap labour to help with both the photography and plate measurement (see *The women of the Carte du Ciel*).

Despite its potential, the project progressed at a glacial pace as observatories struggled with the costs, workload and subsequent devastating effects of the First World War. The photographic atlas was never completed although Greenwich successfully published six volumes of the *Astrographic Catalogue* between 1904 and 1932. Nonetheless, the project demonstrated the value of multinational collaboration and inspired the foundation of the International Astronomical Union in 1919. More recently, astronomers have started to reanalyse the surviving glass plates of the Carte du Ciel to detect the changing positions of the stars (proper motion) over the past century. The lens from the Astrographic Refractor at Greenwich was used on the 1919 eclipse expedition to prove Einstein's Theory of General Relativity (87) and the complete telescope was later installed on new mounting at Herstmonceux (see *Epilogue*). The original mount seen in the photograph remains in the collections at Greenwich today.

⬅
13-inch Astrographic Telescope, Sir Howard Grubb, 1890, in *Greenwich Astrographic Catalogue*, unknown photographer, 1904 (PBC1694/1)

➡
View of the Astrographic Dome (right) from the roof of Flamsteed House, unknown photographer, about 1925 (AST1113.4)

IN FOCUS

# The women of the Carte du Ciel

The creation of the Carte du Ciel ('map of the sky') project in 1887 (71) ushered in both a new age of international collaboration in astronomy and also provided opportunities for women to become professional astronomers, both at Greenwich and elsewhere. For most female participants, it was just another routine office job, but for others it was a useful springboard that helped them develop more sustained and pioneering careers in astronomy and related sciences.

## The task
At the Carte du Ciel's foundational conference, Paris Observatory Director Ernest Mouchez outlined the project's objective of creating a photographic atlas and a catalogue of star positions and magnitudes (brightness) for several million stars. The observers would be required to take photographs in pairs: one long exposure (40 minutes) to create the atlas view and three short exposures (20 seconds to 6 minutes) to be used for measuring star positions. The plates were designed to cover a section of sky two degrees square wide (equivalent to four times the diameter of the Full Moon) with overlapping boundaries to ensure complete coverage. Each plate was to be covered with a grid (*réseau*) of lines 5mm apart to aid the precise measurement of star positions using a microscope once developed. If successful, the project would enable astronomers to map all stars in both hemispheres down to 14th magnitude and catalogued to 11th magnitude, in comparison to the 6th magnitude limit visible to the naked eye.

## The rationale for hiring women
Mouchez and his peers believed that photography would enable them to catalogue more stars in less time with greater accuracy. However, it took much more time to measure and record the plates afterwards than expected. Given that a single observer could now capture multiple photographs in one night, observatory directors realised that they would need several human computers to complete their designated section. Some delegates argued that the plates should be measured centrally in Paris for expediency, while others felt that it should be done at source. Few observatories could afford the additional labour costs and so in August 1891 Mouchez, and his colleagues offered a solution: the recruitment of women.

Perceived to be more patient and conscientious in undertaking repetitive and precise work than their male counterparts, female computers were regarded as the ideal source of cheap labour for this task. Edward Pickering, Director of Harvard College Observatory, had already adopted this approach a few years earlier by employing a group of women astronomical computers to analyse stellar spectra (see *Splitting starlight with spectroscopy*). By 1892, Mouchez had established his own Bureau des mesures at Paris Observatory, staffed by five women and led by the American astronomer Dorothea Klumpke. Collectively, they measured around 30,000 stars each year.

The painstaking work was undertaken in an industrial fashion with a strict regime that included cross-checks for detecting errors and monthly targets to keep on schedule, all

↑
Sisters of the Holy Child Mary community working on Carte du Ciel plate measurement, unknown photographer and date
The Vatican Observatory

overseen by a supervisor in a space segregated from male colleagues to avoid any distraction. The women were generally recruited from previous roles as teachers, telephone operators or post office staff and could be trained in how to measure the plates within a few months. They relied on a purpose-built machine called a 'macromicrometer' in which the photographic plate – illuminated from underneath by reflected light – could be inspected via a moveable eyepiece that featured crosshairs for measuring the coordinates of each star. The women were required to measure around 40 stars per hour and were also responsible for assembling the results for printing, although they were rarely acknowledged in the resulting publications. Similar groups were established at observatories in Toulouse, Bordeaux and Algiers, while four nuns were recruited to work on the Vatican Observatory's contribution in 1910. At Mexico City Observatory, the astronomers enlisted their daughters as computers.

**Christie's 'lady computers' at Greenwich**
Christie's group of four 'lady computers' began work at Greenwich in September 1890. They had all completed degree-level studies in mathematics at Newnham or Girton College, Cambridge University, but were paid at the lowly rate given to the teenage computer boys, rather than the Assistant rate paid to other graduates. Isabella Clemes, Harriet Furniss and Edith Rix only stayed for one or two years but Alice Everett continued until June 1895. In September 1891 she

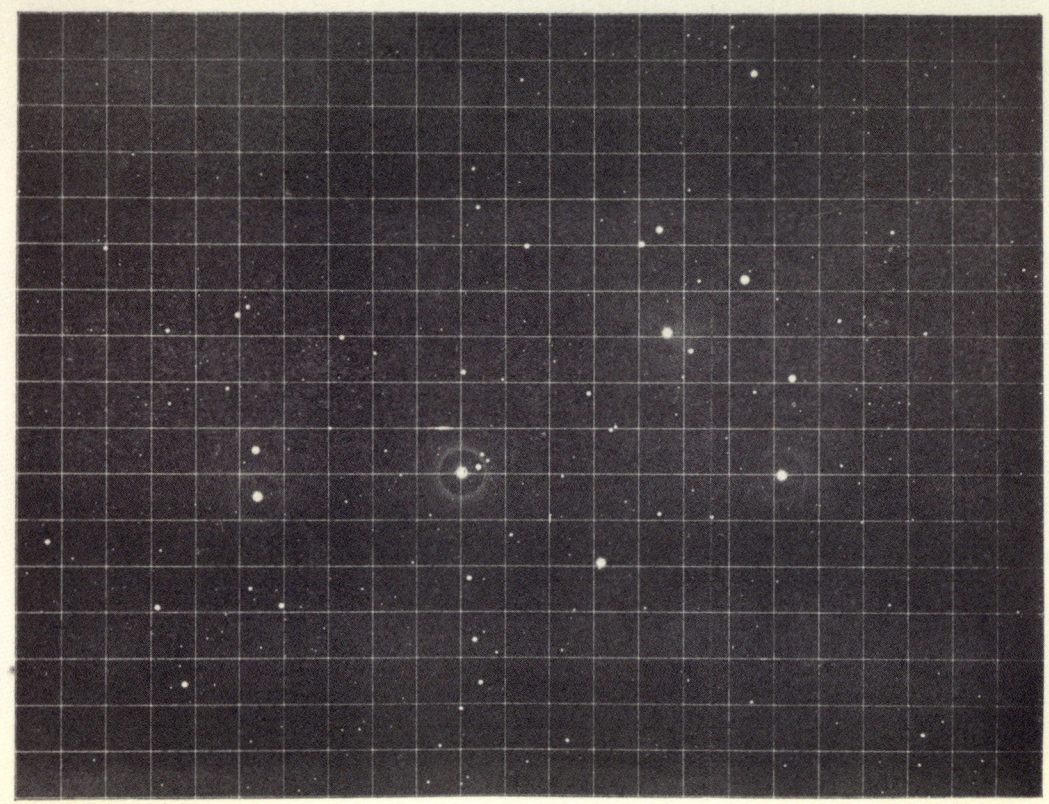

THE STARS OF THE PLEIADES.
(*Photographed with the 'Astrographic Telescope' of the Royal Observatory, Greenwich. Exposure, about 40 minutes.*)

The Stars of the Pleiades print,
E. Walter Maunder in *The Heavens and their Story*, 1908 (PBH8641)

was joined by her former Girton classmate Annie Scott Dill Russell (72, 81), who would remain associated with the Observatory for several decades.

Unlike the desk-bound duties of the women at other observatories, the Greenwich women were offered a reasonable mix of office and observational tasks, as described by Everett herself in a magazine interview: 'The hours are 9–1 every week day; and 2 to 4½ on 3 afternoons a week; with observing on 3 nights, the length of a night's work depending of course on the weather, but not exceeding 3 to 4 hours.' The observing notebooks for the Astrographic Telescope for 1893 show that Everett and Russell captured around six star-fields each clear night and dutifully noted the ambient temperature, pressure and type of photographic plates used. Sometimes they also observed the Moon and planets, or else temporary events such as the bright star (nova) seen in the constellation of Auriga in March 1893 and Comet Quénisset later that summer. The tedious daytime task of measuring star positions was accelerated in January 1895 with the introduction of a machine called a duplex micrometer that enabled them to measure two plates simultaneously using a moveable eyepiece containing crosshairs, Yet, it was still a demanding job that continued for decades (71).

In addition to their Carte du Ciel responsibilities – and in contrast to women computers elsewhere – Christie's longest-serving female employees also had the opportunity to

*Alice Everett*, Morgan and Kidd, around 1893 (ZBB0624)

Duplex micrometer in *Greenwich Astrographic Catalogue*, unknown photographer, 1904 (PBC1694/1)

gain valuable experience on other instruments: Everett made observations with the Airy Transit Circle (47), while Russell took daily photographs of the Sun using a photoheliograph (64) purchased for the transit of Venus in 1874.

**Life after the Carte du Ciel**
Armed with several years' experience of observations and plate measurements, Everett left the Observatory in July 1895 to continue with similar Carte du Ciel work at Potsdam Observatory. A few years later she returned to London and became interested in optics, briefly working for the instrument makers Hilger (67, 96) before joining the Physics Section of the National Physical Laboratory at Teddington, south London, in 1917. She retired in 1925 and pursued a successful third career in wireless and television technologies until her death in 1948.

As required by Civil Service rules, Russell resigned from her role in October 1895 to marry her colleague E. Walter Maunder but she continued to work as a volunteer on sustaining the Observatory's daily photographic sunspot record. Christie ensured that her name was included in official publications and the Maunders subsequently created the so-called 'butterfly diagram' of sunspot activity in 1904 (72). Annie also used her photographic expertise to capture pioneering photographs of the Sun on eclipse expeditions (81) and undertook research on the origins of the constellations. The couple later returned to work at the Observatory during the staff shortage of the First World War (84).

## SOLAR PLATE MICROMETER

Christie originally intended for his 'lady computers' to work on the Carte du Ciel project (71) but the women were gradually assigned to other departments as required. On 4 September 1892, the initials 'AR' appeared in the Observatory's record of solar observations, marking the start of Annie Scott Dill Russell's long association with the subject.

Reporting to E. Walter Maunder, Russell used the Dallmeyer photoheliograph (64) to capture daily images of the Sun. The instrument was modified to produce a larger 8in. (20cm) image in April 1884 and so Christie commissioned this enlarged solar plate micrometer to help with the sunspot analysis. First, the image of the Sun was centred on the plate and the scale set to zero. The astronomer could then measure the distance of a sunspot using the vernier scale down to an accuracy of 0.004in. (0.01cm). The sunspots were measured in terms of their solar latitude and longitude, with corrections made for optical distortion and refraction.

Annie and Walter's working relationship blossomed into friendship and then romance. The couple married on 28 December 1895; for Annie this signalled the end of her professional career, as married women were not permitted to work in the Civil Service at the time. Nonetheless, she continued her observations on a voluntary basis and the Maunders published a range of papers on the number, duration, size and distribution of sunspots, the rotational period of the Sun and the relationship between solar activity and magnetic storms (53). In 1904, they published a paper based on 30 years' worth of Greenwich observations to show how sunspots appear at changing latitudes over the 11-year solar cycle, creating a distinctive butterfly shaped pattern. The Maunders also gave public lectures and wrote popular astronomy books (81) to promote their shared love of the subject.

After returning to the Observatory to assist during the First World War, Walter retired for the second and final time on 30 September 1919. He was succeeded by Junior Assistant H.W. Newton who sustained the daily heliographic observations and sunspot analysis for another 37 years, despite the Second World War and the Observatory's move to Herstmonceux (see *Epilogue*).

←
Mr H.W. Newton, Assistant-in-Charge, examining a solar negative taken by the photoheliograph before putting it in the solar micrometer for the measurement of sunspots, unknown photographer, about 1945 (REG18/000454.37)

→
Solar plate micrometer, Troughton and Simms, 1884 (ZBA0781)

## 73
### 'GREETINGS FROM GREENWICH' POSTCARD

On 20 September 1907, a young girl called Maud sent this postcard to convey birthday wishes to her cousin Winnie in Portsmouth. The friendly chit-chat on the reverse makes no mention of the Observatory but the site is prominently featured among other well-known Greenwich locations, accompanied by celestial motifs. Curiously, the photograph of the Observatory must have been taken over two decades earlier, when the Equatorial Building was topped by a drum-shaped dome, rather than the onion-shaped one installed in 1893 to accommodate a larger telescope (75).

Despite the proliferation of Observatory scenes on postcards such as this, the site was firmly closed to visitors, as described by journalist Philip Astor in 1901: '...the place is dedicated to minute observations and elaborate calculations, which call for a total absence of all disturbing elements'. From the earliest days, visitors were limited to those permitted by the Astronomer Royal. Letters written by Maskelyne's assistants reveal they could only arrange private social visits when their employer was away. Airy also contended with unofficial visits from domestic maids' courting young men or local boys stealing apples from his kitchen garden.

Public visits were a rare occurrence until the 1860s when around 200 guests of the Board of Visitors were invited to Visitation Day to view instruments and ask questions, fuelled by hot chocolate, lemonade and cake. These annual events were strictly all-male until 1880. A few decades later, Astronomer Royal Frank Dyson turned visits into wartime relief with the biennial Observatory Garden Party set-up to raise funds for local hospitals. These popular charity events continued through the 1920s and were supplemented by concerts for wounded soldiers in the Octagon Room (2).

'Greetings from Greenwich', Chaucer Postcard Publishing Co., about 1890
Graham Dolan

## 74
## POCKET CHRONOMETER 'ARNOLD 485'

Ruth Belville with the 'time-keeper' of the South Metropolitan Gas Company who is checking the time from Arnold in her right hand, unknown photographer, 1929
Graham Dolan

'Arnold 485' pocket chronometer, John Arnold, about 1794
The Clockmakers' Museum

Affectionately known as 'Arnold', this pocket chronometer was the linchpin of a family business associated with the Observatory for over a century. Keen to focus on astronomy, Airy was frustrated by the time assistants spent dealing with chronometer makers checking the time. In June 1836, assistant John Henry Belville (51) launched a side business carrying a chronometer, checked against Observatory clocks, to a network of 200 subscribers. When Belville died in 1856, his widow Maria petitioned Airy to continue the service, having no entitlement to her husband's pension. Now reduced to 67 subscribers, Maria's business competed with the new electric time signals from the Shepherd Motor Clock (49), distributed via telegraph.

In June 1892, her daughter Ruth took over. Familiar with the route since childhood, she began each Monday at the Observatory, where the chronometer was checked and certified. After tea with the gate porter, she crossed the city to visit around 40 businesses. Though telegraph companies offered hourly signals, Ruth's clients valued the chronometer's consistency and accuracy to a tenth of a second, unlike the telegraph, which often failed or was delayed. She retired in autumn 1940, as Blitz bombing raids made her route too dangerous.

Originally made in gold for HRH Duke of Sussex, the chronometer was refitted in 1840 with a new escapement (see *How does a mechanical clock work?*) and a plain silver case to avoid attention when travelling through London's slums. As Ruth wished, it was bequeathed to the Worshipful Company of Clockmakers and is now displayed at the Science Museum, London.

## 28IN. GREAT EQUATORIAL TELESCOPE

With the Royal Observatory firmly established in the global scientific community, thanks to the Prime Meridian (69) and its involvement with the Carte du Ciel (71), Christie was keen to ensure that Greenwich maintained its standing. Expressing his fears to the Admiralty in 1885, Christie claimed that the Observatory's 12.8in. (32.5cm) Merz refractor from 1859 was 'very far inferior to those of the chief observatories of Europe and America'. The Admiralty concurred and granted him the funds to purchase a new telescope that would fulfil his ambitious plans.

To keep costs down, Christie ordered the largest possible telescope that could be installed on the existing equatorial mount and still used in conjunction with Airy's water-powered drive to keep in sync with the stars. He commissioned a refracting telescope from the well-known manufacturer, Howard Grubb of Dublin, who subcontracted the glass specialists Chance Brothers of Birmingham to provide the two-part crown and flint lens that eventually weighed over 270lb (122kg).

Christie also planned to reuse the Merz telescope's wooden drum-shaped dome but the structure had deteriorated since the 1850s and so he decided to design a new dome himself. Drawing inspiration from mathematical shapes, he created an onion-shaped profile that could accommodate the larger telescope – only just – within its bulbous walls. The telescope finally came into service in January 1894 but the astronomers struggled with the cumbersome task of changing the configuration of the heavy lenses for different types of observation. The anticipated programme of innovative photographic and spectroscopic observations faltered within a few years and was reassigned to the newer Thompson Reflector (82). The Great Equatorial Telescope was repurposed for observations of double stars (86), planets and comets and remained in use for another four decades.

As the events of Second World War began to unfold in September 1939, the telescope lens was removed for safekeeping. This proved to be a wise decision when the dome's papier-mâché covering was damaged by a nearby bomb blast in July 1944. After the war, the telescope was relocated to Herstmonceux (see *Epilogue*) but was returned to Greenwich as a museum artefact in 1971.

← Great Equatorial Telescope, 28in. (71cm) aperture, Howard Grubb, 1893, installed on an equatorial mount, Ransomes and Sims, 1859 (AST0932)

→ Detailed view of the modern eyepiece installed for public viewing events

(overleaf)
Thomas Lewis and William Bowyer observing double stars with the Great Equatorial Telescope, attributed to David J.R. Edney, about 1900 (B5698-C)

## THE SECRET AGENT (1907)

As the light began to fade during the afternoon of 15 February 1894, the Observatory's assistants William Thackery and Henry Hollis were busy working in the Computing Room when they were startled by an explosion, 'the detonation of which was sharp and clear', followed by a noise 'like a shell going through the air'. Accompanied by porter William McManus they raced towards a plume of black smoke situated on the zig-zag path that approached the Observatory from the west. The trio found the kneeling figure of Martial Bourdin, a 26-year-old French tailor and known anarchist, who had lost his left hand and was bleeding profusely from abdominal injuries. The man was taken to the Seaman's Hospital at the Naval College but died within half an hour. The Observatory staff undertook the grim task of extracting various pieces of skin, bone and bloodied fabric from the surrounding area.

Within hours, the event had been reported in gory detail as the newspapers tried to make sense of the 'Greenwich outrage': was this an attack on the Observatory itself, or was the acid-detonated bomb intended for elsewhere? Did Bourdin just stumble or was he on his way to Paris, which had been the scene of another anarchist attack just days earlier? Others pointed to the significance of the Observatory's gate clock as a public symbol of Britain's imperial dominance of global time and longitude. Bourdin himself uttered no words before his death, leaving behind a mystery that has since been immortalised in various novels and films, most notably Joseph Conrad's *The Secret Agent* (1907) and Alfred Hitchcock's *Sabotage* (1936).

Among Observatory staff, the bomb attack was quietly forgotten. The Astronomer Royal Christie had already left that day for a long weekend away on his yacht and made no reference to the event in his *Annual Report* that summer. Nearly 20 years later, the Admiralty arranged for a police guard at the Observatory to prevent another possible attack, this time from suffragettes, some of whom had targeted the Royal Observatory in Edinburgh on 21 May 1913. Thankfully, none occurred.

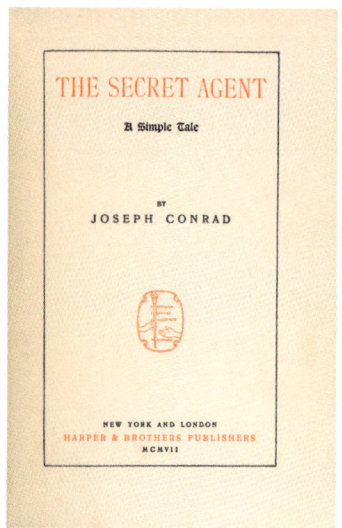

◗ Cover and title page of *The Secret Agent*, Joseph Conrad, Methuen & Co, 1907

# THE
# SECRET
# AGENT

## JOSEPH CONRAD

## 77
## ALTAZIMUTH PAVILION

At the Annual Visitation in June 1892, Christie explained to the Board of Visitors that he needed a new instrument to make additional observations of the Sun, Moon and planets. With its approval, the Astronomer Royal secured the necessary funding from the Admiralty and work began in 1894.

Located at the mid-point of the Observatory site, the new Altazimuth Pavilion was designed by the architect William Crisp, who was also working on the neighbouring New Physical Observatory (78). The name 'Altazimuth' referred to the motion of the Pavilion's 8in. (20cm) aperture telescope in both the horizontal (azimuth) and vertical (altitude) planes (see *Glossary*). Made locally in Charlton by Troughton and Simms, the telescope was supported by a central column and was aligned (collimated) with extra telescopes north and south. The dome was an unusual design with two hemispheres that separated and slid outwards to provide a 4.5ft (1.4m) opening.

After some initial problems, the telescope came into regular use in February 1899 with Andrew Crommelin assigned to oversee the instrument's work. In 1910, astronomers in Germany offered a modest financial reward to anyone who could successfully predict the date when Halley's Comet – due to return imminently – would make its closest approach to the Sun (perihelion). Crommelin and his colleague Philip Cowell submitted their entry and the icy body was eventually spotted on 12 September 1909, providing astronomers with a known position. When the competition entries were unsealed, Crommelin and Cowell's prediction of a close approach on 16 April 1910 was found to be the most accurate: Greenwich had won! Somewhat prophetically, a gilded weathervane shaped like the historic comet had been added to the dome in June 1901.

The Altazimuth remained in use until 1929, despite the vibrations and smoke emanating from the newly built Greenwich power station situated half a mile (0.85km) away. The telescope was scrapped just months before the building was damaged during the Blitz on 21–22 October 1940. Now restored, the Altazimuth Pavilion is home to the modern computerised Annie Maunder Astrographic Telescope (AMAT), named after the pioneering astrophotographer and Greenwich 'lady computer' from the 1890s (72, 81).

The Altazimuth Pavilion with its dome open and telescopes just visible, unknown photographer, about 1905 (A8528-013)

Andrew Crommelin, Elliott and Fry, unknown date (REG18/000454.56)

## NEW PHYSICAL OBSERVATORY

Constructed over eight years in the 1890s, this imposing red brick and terracotta cruciform building marked the peak of Christie's battle to sustain the Observatory's scientific standing and national prestige. While Airy had kept the Observatory's focus on positional astronomy for navigation and timekeeping, Christie, a moderniser, wanted to pursue research opportunities afforded by astrophysics, for which he needed more staff, offices, workshops, dark rooms and storage.

His 1889 request to the Admiralty was modest: a single-storey brick store for moveable and historic instruments to replace shabby wooden huts in the south grounds. Funds were granted and William Crisp was appointed architect. Within a year, Christie's vision expanded, possibly inspired by his summer visit to Pulkovo Observatory near Saint Petersburg with its many wings, domes and a central octagonal tower. After further petitions he secured his 'Physical Observatory', blending function, display and observation with a dome housing the two Thompson photographic telescopes (82). National pride was emphasised by 24 terracotta plaques naming key figures who contributed to the Observatory's legacy. Portholes around the dome and a ship-shaped weathervane echoed ties to the Admiralty. The building also featured its own innovative power supply and warm air heating.

With space limited Christie had to compromise. As the wings took shape in the early 1890s, staff in the nearby Magnet House found the steel frame disrupted the sensitive instruments. Christie secured a parkland site 1,300ft (400m) away for a new Magnetic Enclosure, used until 1925 when the department moved to Abinger, Surrey (91). The Physical Observatory remained in use until the late 1950s when the Thompson telescopes moved to Herstmonceux (see *Epilogue*) where they remain.

◀ The Astronomer Royal's office in the north wing, complete with a display of eclipse photographs and globes, unknown photographer, 1898
(PBC0878)

▶ View of the building with the Thompson Refractor just visible in the dome, unknown photographer, 1900
(P39986)

IN FOCUS

# Sports and social clubs

On Saturday 21 April 1881, Chief Assistant H.H. Turner noted in his journal that the Observatory's staff had spent the afternoon enjoying an idiosyncratic sports day which seems to have been a unique event in the institution's history. Competitors took part in an anemometer race, a tug of war and a race to rotate one of the domes in the shortest time possible, an event won by Turner himself in just 30 seconds. It was the start of a wider social trend in the late 1800s in which schools and universities fostered a sense of community and camaraderie through competitive sports. As the Observatory began to recruit more graduates through the Civil Service Entrance Examination, the assistants took inspiration from their educational experiences to create a range of workplace sports and social clubs. Despite the strict social hierarchy between the temporary computers and permanent assistants, the clubs were well attended and flourished until the disruptive years of the First World War. After the Observatory moved to Herstmonceux in 1948 (see *Epilogue*), the astronomers took advantage of the sprawling countryside location to invest in a much wider range of leisure activities. This provision cultivated a strong sense of community and identity that surviving members still fondly recall today.

### The Hockey Club
While the Cricket Club at Greenwich only lasted for two years (1892–94), the Hockey Club started in 1893 and endured for over 40 years. Turner was the first captain and within a few seasons the team had adopted its colours of a navy-blue shirt with scarlet collars and cuffs. Participants were expected to attend mid-week practice and the Astronomer Royal permitted them to finish work early during the winter months to play during the last few hours of daylight. The initial sessions were held on a field in Kidbrooke, south-east London, but over the years the Hockey Club eventually migrated between fields in Blackheath and Lewisham before settling on the closer Ranger's Meadow in Greenwich Park. Matches were held from September to April and the Observatory team played against others from local factories, railway companies, hospitals and schools. There was even a strong sense of competition within the team itself with an annual tournament of Staff versus Computers. Around 1900, Captain Philip Cowell organised a Boxing Day match against Blackheath Ladies' Hockey Club – assembled by his wife Phyllis – but the Ladies were heavily defeated and no further mixed games took place. At the end of the season, the players celebrated with an annual supper held at the riverside Trafalgar Hotel – complete with songs on the pianoforte – and sometimes enjoyed card games such as 'whist' during the winter months.

Reminiscing decades later in the 1950s, Assistant William Witchell recalled how his colleagues each played hockey in their own unique style, from Bryant's avoidance of wearing

Probably a group photo from one of the Camera Club's excursions, unknown photographer, about 1900 (AST0050.134)

The Hockey Club, unknown photographer, 1890s (REG18/000454.75)

shin protectors ('and suffered accordingly') to Burkett's powerful attack and Evans' deadly shots from seemingly impossible angles. With a sense of wistful pride, Witchell concluded that the 'grand winter game' contributed to the prestige of the 'ancient institution'.

## The Camera Club

For those less interested in sports, the Observatory's Camera Club offered staff members the opportunity to hone their skills in social photography rather than astrophotography. Astronomer Royal Christie had already gained over 20 years' experience of photography by the time he became inaugural President of the Blackheath Camera Club (BCC) in 1891. Open to both amateur and professional photographers, the BCC attracted over 75 members – including a handful of women – and held a two-day exhibition in a local hall that was attended by over 600 people. The invention and mass production of gelatine dry plates in the 1870s had made photography more affordable and accessible, leading the number of clubs in Britain to swell from 14 in 1877 to over 365 by 1910. They offered members a sociable forum where they could seek advice or share expertise, listen to lectures, use dark rooms for processing and participate in excursions and competitions.

The Observatory's own camera club came into existence in 1900, possibly in response to a competition organised by H.H. Turner, by then working as Director of the University Observatory, Oxford. Turner was due to give a public lecture in February 1900 and so, a few months in advance, he challenged his former colleagues to send him the best set of slides that illustrated 'the Old Royal Observatory and the New'. The competition was open to any member of staff and offered a significant financial reward to successful entrants. Sadly, we don't know if anyone actually took up the challenge but, a year later, the Observatory's fledgling Camera Club held its first annual exhibition in the North Library (since demolished), starting a trend that continued until 1906. The exhibition initially had five categories (architecture, portraiture, landscape, seascapes and enlargements) and was accompanied by items on loan from the Royal Photographic Society. Staff members D. Edney, W. Bowyer, P. Melotte, W. Stevens and H. Furner dominated both the entries and awards each year, a trend that perhaps contributed to the addition of new categories in 1906, including one for 'members who have never obtained a prize'. Members of the Camera Club also started to take advantage of local railway connections to make excursions into Kent, Surrey and Essex where they took prize-winning photographs of Ightham Mote, Richmond Park and other landmarks. The last exhibition was held in 1906, after which the official records cease.

## 79
## SPIDER FORK

Spiders have long had a duplicitous reputation at the Observatory, regarded as both friend and foe. Working in the 1600s, English astronomer William Gascoigne was apparently inspired by a wayward arachnid spinning a web across his telescope when he suggested that spider silks could be used as vertical lines in the field of view – similar to rifle crosshairs – to help observers measure the crucial moment of transit. Over the next two centuries, astronomers experimented with various materials, concluding that spider threads were stronger, thinner and more uniform than silk, human hair, or metal wires.

By the 1800s, supplying spider threads as 'reticules' for telescopes was part of everyday business for instrument makers. Replacement threads for Greenwich instruments were ordered from Troughton and Simms and the 1864 inventory notes 'a box of cobwebs' kept alongside the Airy Transit Circle (47). This box likely contained forks similar to the one shown, on which spiders were enticed to spin by rotating the gadget. About a dozen threads could then be kept taut or extracted as needed. This homemade method was feasible in Greenwich Park but less so overseas, where astronomers faced potentially poisonous spiders while maintaining telescope and theodolite accuracy.

Even harmless local spiders were not always welcome. When journalist Frederick Knight Hunt visited in 1850, he described how Airy's assistants struggled to keep a telescope tube clean because its 'cool and dark interior was so pleasant to the spiders that, do what they would, the astronomers could not altogether banish the persevering insects from it'. But, as Hunt observed, arachnids venturing onto electric instruments in the Magnet House were less fortunate: 'The spider who had ventured on the charged wire paid the penalty of such daring with his life'.

Metal fork, made in the Observatory's workshop, unknown maker, 1900–50 (ZBA9426)

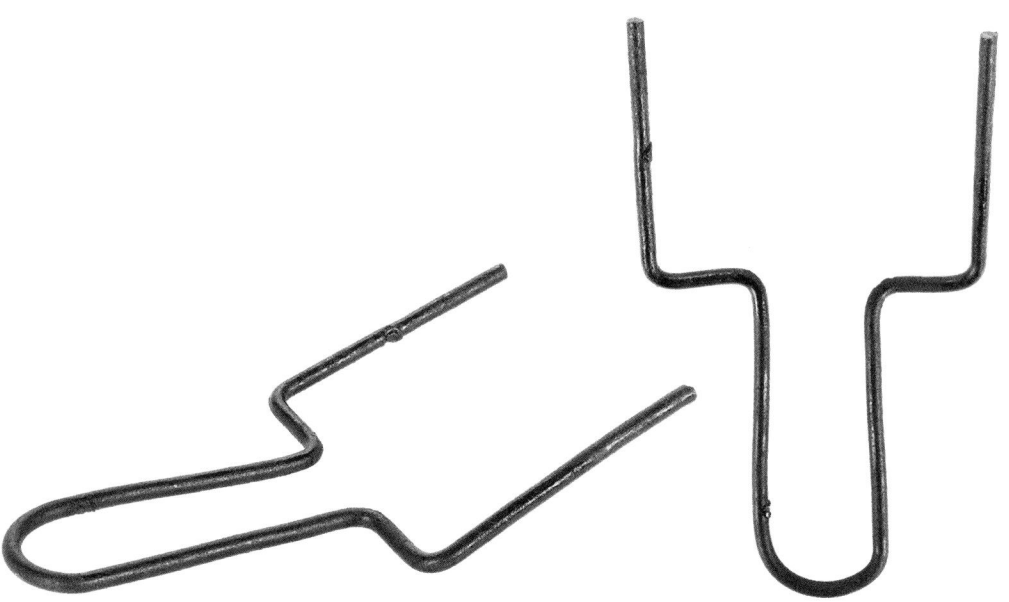

214  A HISTORY IN OBJECTS

## ARTICLE BY MAUNDER ABOUT THE 'CANALS' ON MARS

At just 14 pages long, this groundbreaking article marked a turning point in the debate about the nature of canal-like structures and the possibility of extra-terrestrial life on Mars. Assistant E. Walter Maunder (64, 72, 81) had been fascinated by the Red Planet throughout his career at Greenwich. In September 1877, just four years into the job, he used a spectroscope (see *Splitting starlight with spectroscopy*) to measure the spectrum of the Martian atmosphere when the planet was favourably situated opposite the Sun and close to Earth (perihelic opposition). At the same time, Giovanni Schiaparelli at Brera Observatory in Milan was compiling a new chart of Martian features, including linear markings known as 'canali'. When translated into English as 'canals', this triggered a decades-long debate on whether they were natural formations or artificial structures made by intelligent Martians. By the late 1880s, Maunder voiced suspicions they were merely 'optical products' but it was this article in 1903 that firmly established him as an opponent to the speculation.

Over a year, Maunder enlisted J.E. Evans, Headmaster of the nearby Royal Hospital School, Greenwich, to recruit teenage boys who had never seen Mars through a telescope for nine visual perception experiments. The boys sat at differing distances from a small reproduction of the Martian surface, which changed each time to show sketches by different astronomers. Armed with pencils and paper, they were asked to sketch what they could see, with no indication of the subject. As expected, those closest included the most detail while those furthest away produced more abstract markings. Maunder showed how the boys unwittingly created lines between individual dots, like Schiaparelli's 'canali'. It was compelling evidence both for resolving the Martian debate and for the broader question of visual perception in astronomy.

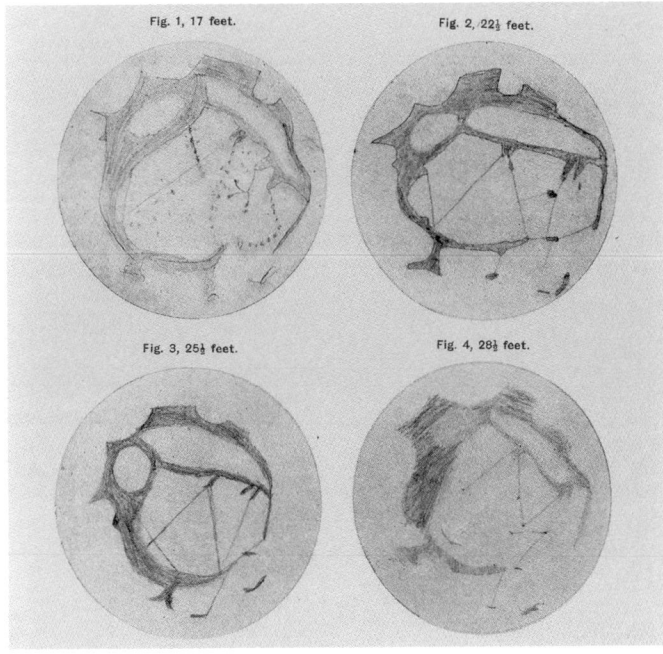

➲ The experimental results as seen from increasing distances, published in *Monthly Notices of the Royal Astronomical Society*, Maunder and Evans, 1903
(MNRAS 1903)

# THE
# HEAVENS
## AND THEIR STORY

A. & W. MAUNDER

## 81

## THE HEAVENS AND THEIR STORY (1908)

Despite having to resign from her role at Greenwich upon her marriage in 1895, Annie Maunder continued to participate actively both in her sunspot research and popular astronomy. Five years previously, her husband Walter had been one of the founding members of the British Astronomical Association (BAA), which unlike the elite Royal Astronomical Society (RAS), was open to all, regardless of gender or professional status. The couple dedicated themselves to organising BAA events such as meetings, public lectures, eclipse expeditions and writing a steady stream of accessible and engaging popular astronomy articles and books. One Greenwich colleague, Henry Park Hollis, commented on how the British Museum Library's worn copy of Walter's *Astronomy without a Telescope* (1904) had 'since been replaced by a new copy, which has not happened to most of the books on the shelf'. Similarly, Walter's *The Stars as Guides to Night Marching* (1916) generated letters of appreciation from soldiers on the Western Front during the First World War.

Published in 1908, *The Heavens and their Story* is the only book issued under both their names, although Walter openly states in the introduction that it is 'almost wholly the work of my wife'. The blue and gold cover with the Milky Way arching above the outline of Flamsteed House clearly demonstrates the authors' association with the Observatory. As first author, Annie is listed as an 'honorary fellow of the Royal Astronomical Society of Canada'; it would take another eight years before she and other women would be admitted as full Fellows of the equivalent society in London.

The book is divided into four sections, encompassing topics such as the apparent motion of the Sun, Moon and stars, the planets ('The Sun's Family'), nebulae and the Milky Way. As one might expect, the section on the Sun is richly detailed and strongly reflects Annie's confidence and expertise in the subject. The book also highlights her proficiency in astrophotography, with two of her most famous eclipse expedition images, namely a wide-angle shot of a large bright structure (coronal streamer) emanating from the solar surface, as seen from India in 1898, followed by an image of the Sun's outer atmosphere (corona), as seen from Mauritius in 1901.

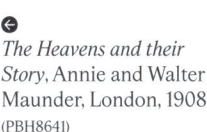

← 
*The Heavens and their Story*, Annie and Walter Maunder, London, 1908
(PBH8641)

→
Annie and Walter Maunder, Daniel Brand Marsh, unknown date
Cambridge University Library

82

## 30IN. THOMPSON REFLECTOR

On 29 January 1908 Jupiter was at opposition, rising above the eastern horizon as the Sun was setting in the west. Established Computer Philibert J. Melotte routinely took advantage of the favourable views and long winter nights to photograph the planet using the Observatory's largest telescope, the 30in. (76cm) Thompson Reflector. As he analysed the photographic plates afterwards, he realised that the large gap between Jupiter's moons hinted at the presence of intervening moons. By the end of February, he realised that he had already captured a new, eighth moon on the plate for 27 January 1908. Melotte's discovery was confirmed by fellow observers at Heidelberg and Lick observatories and the new Jovian satellite was later named after the Greek goddess Pasiphae.

For Melotte, the discovery was just one of many achievements during his 53-year career at Greenwich that saw him progress from teenage computer to astrographic expert. For the Astronomer Royal, the discovery of Jupiter VIII vindicated his earlier decision in 1894 to accept the reflector endowed by the wealthy London surgeon, Sir Henry Thompson. The reflector was made by Howard Grubb of Dublin, although the mirror itself was designed and made by the famous amateur astrophotographer Ainslie Common. The instrument was installed in the dome of the New Physical Observatory (78) and was counterbalanced by another 26in. (66cm) refractor by Grubb that was later used for photographing the Sun, planets and star fields. Christie opted to install both photographic telescopes on a clockwork-driven German-style equatorial mount to enable continuous movement for long exposures.

The reflector became the Observatory's workhorse for studying the Solar System with photographs taken of Halley's Comet (77) and the moons of Mars, Jupiter, Saturn and Neptune. In the 1920s, the telescope was briefly used for measuring stellar temperature and spectroscopy (see *Splitting starlight with spectroscopy*). By the 1930s, however, the Thompson Reflector was superseded by another donated instrument, the 36in. (91cm) Yapp Telescope with its dedicated spectrograph (96). As war broke out in September 1939, key components from the Thompson telescopes were removed for safekeeping and both instruments were later transferred to the Observatory's new location in Herstmonceux.

➔
Philibert J. Melotte examining photographic plates, unknown photographer, 1933
(AST1113.20)

➔➔
Postcard showing the Thompson Refractor (left) and Reflector (right) telescopes by Howard Grubb, Henry Richardson, about 1908
Graham Dolan

ROYAL OBSERVATORY, GREENWICH. SERIES. No 8.

# Frank Watson Dyson

## 1868–1939

Born in Derbyshire and raised in Yorkshire, Frank Dyson was a prize-winning Cambridge mathematician who was appointed as the Observatory's Chief Assistant in March 1894, a few weeks after the anarchist bomb attack. His first task was to assemble the Greenwich data for the International Astrographic Catalogue, followed by detailed work on the long-term movement of the stars. Twelve years later, Dyson, his wife Caroline and their six children relocated to Edinburgh for his new appointment as Astronomer Royal for Scotland, but within a few years, Christie had retired and Dyson returned to Greenwich as his successor in 1910.

With the outbreak of the First World War, the ninth Astronomer Royal and his family stoically remained on site, enduring the Zeppelin raids that forced them to shelter in the cellars of Flamsteed House. Despite losing the majority of his human computers to military service, Dyson kept the Observatory functioning throughout the conflict. He even took on additional secret war work and welcomed refugees from Belgium. After the war, Dyson promoted collaboration in science through two key projects: firstly, the eclipse expedition of 1919 to test Einstein's Theory of General Relativity, which was successful, and secondly, the creation of the International Astronomical Union, which continues to underpin professional astronomy.

In the 1920s, Dyson and his staff sought to improve the Observatory's measurement and distribution of time standards, starting with the installation of the Shortt free-pendulum clock system, followed by collaboration with the BBC in February 1924 to transmit the national 'six pips' time signal. From December 1927, a similar GMT time signal was issued to ships across the globe via the powerful Rugby radio transmitter. Dyson also supported Rupert Gould's work in preserving the Harrison timekeepers.

With a Baptist heritage and strong Christian beliefs, Dyson was an active member of the parish church, St Alfege, and a keen advocate for organising Observatory events to fundraise for local charities. As he approached retirement in 1933, the wealthy businessman William J. Yapp donated a 36in. (91cm) reflecting telescope to the Observatory in honour of Dyson's achievements. After a busy few years of writing and travelling, Dyson died on board ship while returning to home to Greenwich from Australia in May 1939. He was buried at sea in the Indian Ocean.

*Sir Frank Dyson*, Ernest Moore, 1915–25 (ZBA0724)

## SPHERICAL PLATE CALCULATOR FOR STAR COORDINATES

This simple but ingenious revolving plastic plate is a powerful calculating instrument that helped early twentieth-century astronomers refine their skills in measuring long-term changes in star positions (proper motion). Scottish civil engineer and publisher Walter Biggar Blaikie was known for annual star charts that he published for amateurs from 1898 to 1920 and which were continued by Greenwich astronomer David Edney until 1940. Around 1910, Blaikie devised this stellar coordinate instrument to make it easier for astronomers to solve calculations based on spherical trigonometry. Possibly having met Blaikie within Edinburgh's intellectual community during his time as Astronomer Royal for Scotland, Dyson promoted the graphical calculator within his scientific papers on proper motion and later exhibited the device at meetings of the British Astronomical Association (81) in 1916 and 1917.

Blaikie's instrument consists of two circular discs (with a diameter of 19.6in. (500mm) and made from celluloid) that rotate around a common centre. The letter 'Z' at the top refers to the observer's zenith (the point directly overhead) while the letters 'H-R' indicates the horizon. Dyson plotted a star's position and then calculated the spherical angle between the star, north celestial pole and the distant point towards which the Solar System appears to be moving with respect to the local stars (known as the solar apex). He then used the historic star coordinates as measured by Bradley in 1755 (31) to determine how much the stars had drifted over the centuries.

Blaikie's device was a timely addition to the ongoing debate among professional astronomers about the position and distribution of the stars. A few years earlier, Dutch astronomer Jacobus Cornelius Kapteyn had used a novel statistical approach to show how proper motion was not random but instead tended towards two opposite directions in space. Dyson and his Chief Assistant, Arthur Eddington investigated Kapteyn's claims, gaining valuable experience in measuring tiny changes in star positions that later helped them prove Einstein's Theory of General Relativity (87). Blaikie's spherical slide rule may have been constructed in a simple fashion but its contribution to the art of measuring star positions would have profound consequences.

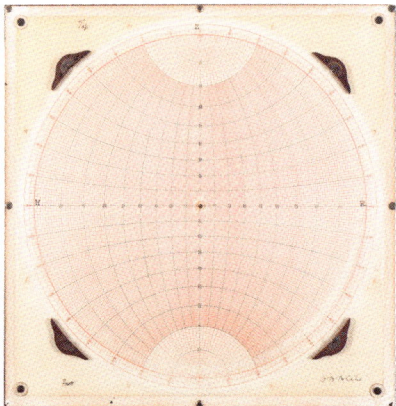

Spherical plate calculator and carry case, Walter Biggar Blaikie, about 1910
(ZBA0775)

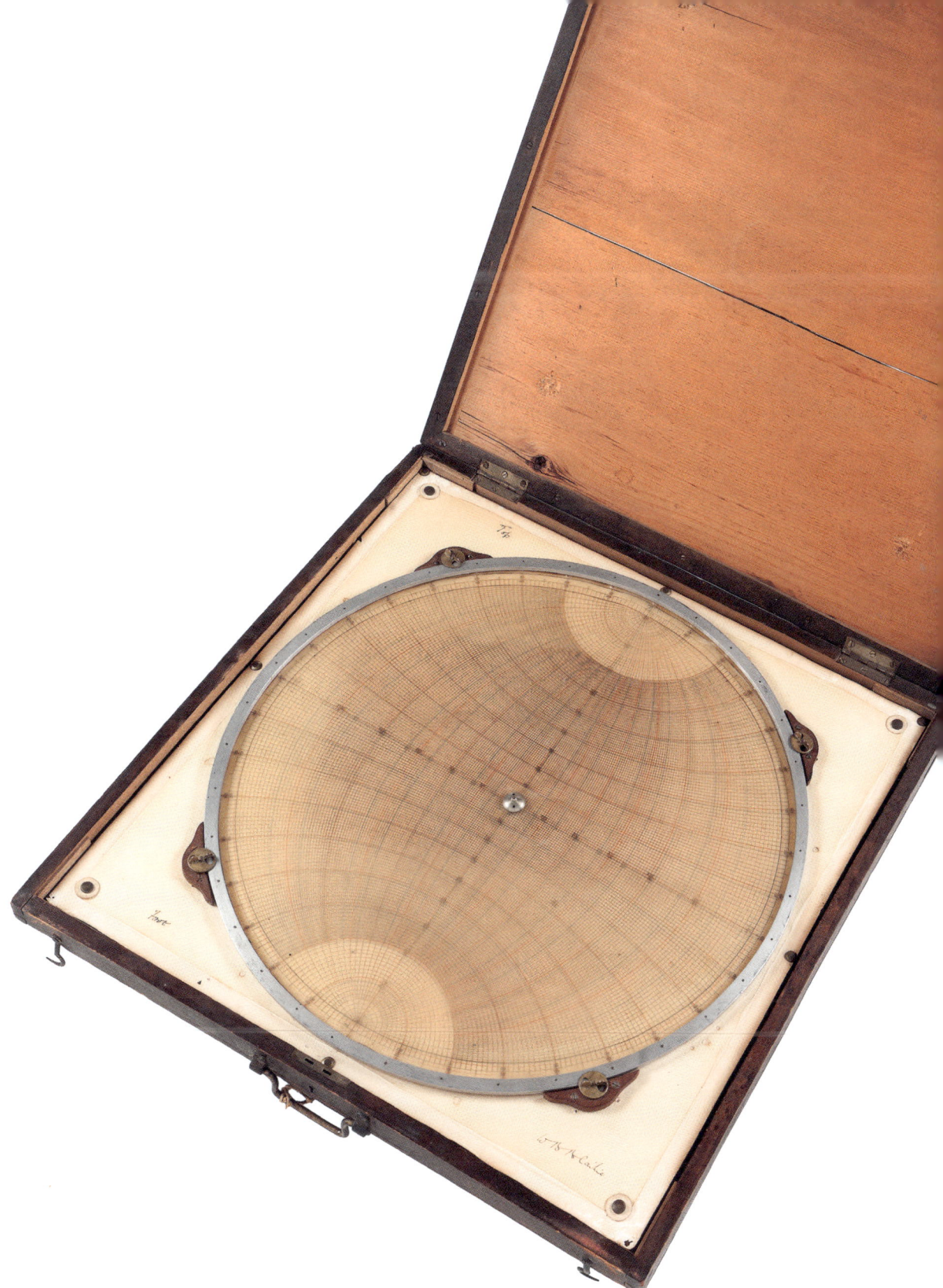

E. J. Adams
46, Vanbrugh Park
Blackheath.

## HENRY OUTHWAITE, OBSERVATORY SECRETARY

In the summer of 1914, Dyson sailed on board SS *Ascanius* to attend a scientific conference in Australia, unaware that he would return a few months later to find Europe at war. Similarly, two of his assistants travelled to Minsk, in modern-day Belarus, to observe the total solar eclipse on 21 August, but the burgeoning hostilities disrupted their plans, forcing them to leave their equipment in haste at Pulkovo Observatory, near Saint Petersburg.

At first, the war made little impact on the Observatory's ongoing work, yet behind the scenes the number of Admiralty chronometers deposited for testing began to increase, along with over 3,000 binoculars for optical testing (85). Within a year, the call for military service had taken its toll as the majority of the Observatory's 26 computers voluntarily signed up. In parallel with the heads of other Civil Service organisations, the Astronomer Royal continued to support the computers' pay and positions, and so, over the next four years, the Observatory's Clerical Assistant, Henry Outhwaite, became an essential point of contact between the organisation and its absent staff. Outhwaite had already been at the Observatory for over 20 years and was responsible for answering letters from the public, managing wages and organising the Annual Visitation. He was too old for active military service himself but two of his assistants responded to the call-up: reservist William Burkett and Horatio Kilby, who tragically died of appendicitis while stationed at Salonika (Thessaloniki, Greece) in 1916.

For Outhwaite himself, the conflict generated a flurry of correspondence as the Computers turned to him for reassurance about their pay and roles. The surviving collection of over 300 letters vividly chronicles the soldiers' evolving wartime experience, from youthful exuberance in the early months to the subsequent grim reality of gas attacks and trench warfare: '[I] am writing this in a rough tent, surrounded by seas of watery mud'. They also reminisce about the happy days of the Hockey Club and their Camera Club expeditions with Outhwaite himself (see *Sports and social clubs*).

While Outhwaite remained at the Observatory throughout the war, other senior assistants were seconded to specialist technical assignments at Woolwich Arsenal, the Royal Engineers and the Hydrographic Office. Outhwaite retired in 1919 and was succeeded by his returning assistant, William Burkett.

*Henry Outhwaite,*
E.J. Adams, about 1910
(REG18/000454.19)

# FIRST WORLD WAR BINOCULARS

Typical of the period, these binoculars remind us of an important but little-known story of the Observatory's contribution to the First World War. In September 1914, just a month after the beginning of the conflict, a new organisation called the National Service League made a public appeal for binoculars as there were insufficient supplies for thousands of new recruits. By 21 January 1915, the League's London office had already received 18,000 instruments but many required cleaning and testing before use. Although some pairs sent from British families living in the tropics were unusable due to corrosion, those confiscated from captured Germans were praised for their high-quality manufacture. Large quantities of binoculars were also imported from the United States but, made hurriedly with poor quality glass, they required reconditioning before issue. With extensive experience servicing optical equipment, the Observatory was ideally suited for such a task and commenced work in September 1915.

Over the next 30 months, a group of 12 Observatory staff worked in teams to carefully dismantle, clean and reassemble a total of 3,050 binoculars. Prior to despatch, the instruments were adjusted and aligned using the image of two illuminated crosses that were directed through a 6in. (15.24cm) telescope and focused by the binocular's prisms onto a screen. Recognising the need for wartime secrecy, Dyson deliberately omitted any mention of binocular testing in his Annual Reports. Our only surviving record of this remarkable project is a single typewritten sheet in the archives, thought to have been composed by Assistant Ernest Martin. As the Observatory prepared to move from Greenwich to Herstmonceux in the late 1940s (see *Epilogue*), Martin discovered the wartime ledgers containing the binocular serial numbers while clearing out cupboards. As the only surviving staff member who had been directly involved, he decided to record the story for posterity and later published his account in the staff magazine.

We have no evidence to link this specific pair of binoculars with the Observatory's testing programme but, like many military optical instruments of the day, they were made by Dollond of London and feature the name 'J.S. Rofe' accompanied by 'R.N.D.' for 'Royal Naval Division'. They were donated to the Museum by Jack Rofe's family, along with his medals, in 1988.

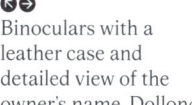

Binoculars with a leather case and detailed view of the owner's name, Dollond, 1915 (NAV0069)

## DOUBLE STAR CATALOGUE BY JONCKHEERE

This simple catalogue of 200 pages filled with numbers may seem mundane but it bears witness to an incredible story of friendship, perseverance and international collaboration in astronomy. Son of a wealthy Belgian industrialist, Robert Jonckheere was an amateur astronomer and skilful observer of double stars who established his own private observatory near Lille, northern France. When the Germans arrived on 3 October 1914, Jonckheere tried in vain to protect the site as his wife and daughters escaped to London. He joined them on 13 October, having walked for three days to reach Boulogne by surviving on morsels of bread and chocolate. Jonckheere was welcomed in Greenwich by the Dysons and became one of seven refugees at the Observatory.

Jonckheere's first task was to observe the transit of Mercury on 7 November 1914 with the Great Equatorial Telescope (75). He also discovered another example of a cloud of gas (planetary nebula) around an ageing star and recorded the return of Wolf's Comet in July 1918. But it was the continuation of his double star work that really gave him the opportunity to excel. Using the Great Equatorial Telescope's precision eyepieces (micrometers), Jonckheere measured the angle between pairs of stars that appeared to orbit a common centre of gravity. William Herschel (30) had first taken an interest in these stars in 1779, questioning whether they were just a line-of-sight effect, or whether their separation could be used to measure the distance to stars (stellar parallax). By Jonckheere's time, astronomers had realised that they could use the separation between double stars to calculate each star's mass and gain insights into stellar evolution. Assembling all his observations since 1905 into this catalogue of 3,950 double stars, the refugee astronomer gratefully acknowledged the Observatory's 'kindness and support', although he admitted that he had struggled with the telescope's 'fictious [fictitious] elongations', seen in certain conditions.

Jonckheere departed from Greenwich on 27 January 1919, returning to Lille to find his family business and observatory in ruins. Undeterred, he continued his double star observations at the observatories of Toulouse, Nice and finally Marseilles, where he died in June 1974.

'Catalogue and Measures of Double Stars', Robert Jonckheere, published in *Memoirs of the Royal Astronomical Society*, 1917

## Mr. R. Jonckheere, Catalogue and Measures of Double Stars.

| No. | Name. | B.D. | R.A. 1920. h m s | Decl. 1920. ° ′ | Angle. ° | Distance. ″ | Magnitudes. | | 1900+ | Obs. | n. |
|---|---|---|---|---|---|---|---|---|---|---|---|
| 912 | J 654 | Anon. | 5 29 47 | 12 45 | 250·0 | 4·00 | 9·2 | 9·4 | 11·87 | J | 1 |
|  |  |  |  |  | 250·5 | 4·28 | 9·3 | 9·6 | 11·87 | V | 1 |
|  |  |  |  |  | 246·0 | 4·38 | 9·3 | 9·7 | 16·10 | J | 1 |
| 913 | Hu 1229 | +37°1242 | 30 0 | 37 51 | 201·6 | 1·74 | 7·5 | 13·0 | 05·32 | Hu | 1 |
|  |  |  |  |  | 198·2 | 1·69 | .. | .. | 05·71 | A | 1 |
| 914* | Bowyer | .. | 30 9 : | 26 52 : | 287·8 | 4·43 | .. | .. | 97·01 | WB | 1 |
| 915 | A 2354 | +18° 881 | 30 17 | 18 35 | 126·1 | 0·50 | 9·5 | 9·5 | 11·75 | A | 2 |
| 916 | J 247 AB | +20°1004 | 30 26 | 20 15 | 145·4 | 4·95 | 9·2 | 9·0 | 10·90 | J | 1 |
|  |  |  |  |  | 146·7 | 4·84 | 9·3 | 9·6 | 16·17 | J | 2 |
|  | AC |  |  |  | 216·8 | 18·75 | 9·3 | 10·8 | 16·17 | J | 2 |
| 917 | J 248 | Anon. | 30 27 | 18 36 | 31·1 | 3·98 | 9·0 | 9·5 | 10·90 | J | 1 |
|  |  |  |  |  | 29·9 | 3·69 | 9·0 | 9·6 | 10·90 | V | 1 |
|  |  |  |  |  | 33·1 | 3·40 | 9·4 | 9·9 | 16·19 | J | 2 |
| 918 | A 2648 | + 5° 959 | 30 32 | 5 25 | 77·4 | 2·80 | 9·0 | 12·8 | 13·96 | A | 2 |
| 919 | A 2107 | +21° 896 | 30 39 | 21 9 | 329·8 | 3·05 | 8·5 | 11·0 | 09·91 | A | 2 |
| 920 | A 1307 | +58° 845 | 30 40 | 58 30 | 131·7 | 4·02 | 8·4 | 12·2 | 06·86 | A | 3 |
| 921 | A 1308 | +59° 896 | 31 0 | 59 47 | 101·6 | 4·82 | 9·0 | 14·5 | 06·89 | A | 3 |
| 922 | Lewis | .. | 31 : | 29 45 : | 197·5 | 0·30 | 10·0 | 11·0 | 08·11 | L | 1 |
| 923 | E 1141 | +48°1290 | 31 7 | 48 11 | 267·9 | 3·39 | 9·1 | 9·6 | 12·18 | E | 3 |
| 924 | A 1562 | +43°1314 | 31 7 | 43 36 | 347·1 | 0·47 | 8·8 | 8·8 | 07·73 | A | 3 |
| 925* | J 676 | + 7° 938 | 31 11 | 7 20 | 100·4 | 1·74 | 8·9 | 8·9 | 11·97 | J | 1 |
|  |  |  |  |  | 98·0 | 1·30 | 9·0 | 9·0 | 11·97 | V | 1 |
|  |  |  |  |  | 102·3 | 1·55 | 9·5 | 9·5 | 13·96 | J | 1 |
|  |  |  |  |  | 100·6 | 1·05 | 8·9 | 8·9 | 16·10 | J | 1 |
| 926 | A 2650 AB | + 8°1011 | 31 42 | 8 4 | 121·4 | 0·57 | 9·4 | 9·9 | 13·96 | A | 2 |
|  | AC |  |  |  | 128·0 | 12·80 | 9·4 | 13·5 | .. | A | .. |
| 927 | A 2510 | + 2°1016 | 32 31 | 2 38 | 278·4 | 0·96 | 8·9 | 11·7 | 13·03 | A | 2 |
| 928* | J 798 | − 6°1253 | 32 32 | − 6 45 | 220± | 3± | 8·9 | 13·0 | 12·15 | J | 1 |
| 929 | A 1563 | +42°1354 | 32 40 | 42 58 | 272·8 | 0·18 | 8·9 | 8·9 | 07·75 | A | 3 |
| 930 | J 147 AB | +23° 976 | 32 58 | 23 18 | 354·4 | 3·45 | 9·5 | 9·5 | 10·71 | J | 1 |
|  |  |  |  |  | 359·2 | 3·40 | 9·4 | 9·4 | 16·10 | J | 1 |
|  | AC |  |  |  | 320·2 | 17·37 | 9·4 | 12·0 | 16·10 | J | 1 |
| 931 | A 2651 AB | + 4° 989 | 32 59 | 4 43 | 158·0 | 0·47 | 8·4 | 10·4 | 13·96 | A | 2 |
|  | AB−C |  |  |  | 325·8 | 8·43 | .. | 12·0 | 13·92 | A | 1 |
| 932 | J 249 | Anon. | 33 8 | 1 13 | 223·0 | 3·00 | 9·2 | 11·0 | 10·93 | J | 1 |
| 933 | J 901 | Anon. | 33 19 | 31 58 | 149·8 | 2·97 | 9·9 | 9·9 | 12·72 | J | 1 |
|  |  |  |  |  | 150·4 | 2·95 | 10·5 | 10·5 | 12·72 | V | 1 |
| 934 | A 2708 | + 8°1019 | 33 31 | 8 54 | 265·2 | 0·46 | 8·5 | 9·3 | 14·73 | A | 2 |
| 935 | A 1564 | +43°1320 | 33 34 | 43 40 | 341·2 | 0·26 | 8·4 | 8·4 | 07·74 | A | 3 |
| 936 | J 250 | Anon. | 33 36 | 1 19 | 167·0 | 3·33 | 9·4 | 13·0 | 10·93 | J | 1 |
| 937 | J 331 | − 6°1266 | 33 56 | − 6 29 | 347·7 | 3·60 | 8·9 | 9·5 | 11·09 | J | 2 |
|  |  |  |  |  | 344·4 | 3·62 | 9·0 | 9·6 | 11·09 | V | 2 |

914—In the field with Σ 749 : 168°9, 0″79, (7·0–7·1), 1904·03, 2.—W.B. The coordinates of Σ 749 are given here.—J.
925—Measured by Aitken as A 2649. In A.G. Leipzig II. the magnitude is 8·7.—J.
928—There is perhaps also a 15th mag. at 40°±5″±. The principal star appears in the centre of the nebula H IV. 33. and is represented as the nucleus of the nebula in the drawing of Rosse, *Trans. Royal Dublin Society*, vol. ii., n.s., plate I.—J.

↑
Glass photopositive of the 1919 eclipse as seen from Sobral, Brazil, on 29 May 1919, with the prominence on the upper right limb, Andrew Crommelin and Charles Davidson, 1919
(AST1091)

## 87

## GLASS PHOTOPOSITIVE OF THE 1919 TOTAL SOLAR ECLIPSE

On 10 November 1917, Dyson convened a meeting to persuade colleagues at the Royal Society and Royal Astronomical Society to fund an expedition to view the eclipse of 29 May 1919, a rare event that offered ideal conditions for testing Einstein's 1915 Theory of General Relativity. Communications between British and German scientists had been curtailed during the war and few people knew about Einstein's work, apart from the Cambridge University astronomer Arthur Eddington, who had been in correspondence with the Dutch mathematician Willem de Sitter.

Despite some anti-German sentiment within the scientific community, Dyson's proposal was accepted and planning began for two expeditions to be led by astronomers from Greenwich and Cambridge: one to Sobral in northern Brazil and one to Principe, an island off the west African coast. Having worked with Eddington at Greenwich, Dyson was keen to involve him, but by January 1918 the Cambridge astronomer was facing prison due to his Quaker beliefs and refusal of military service. Dyson petitioned on his behalf and Eddington was spared, finally setting sail in March 1919 and arriving at Principe just one month before eclipse day.

Despite battling against the clouds, mosquitoes and mischievous monkeys who tampered with the equipment, Eddington and his clockmaker colleague Edwin Cottingham managed to take 16 photographic plates of the eclipsed Sun against the backdrop of the Hyades star cluster. The Sobral team of Andrew Crommelin and Charles Davidson took 19 plates with an astrographic telescope (71) and successfully photographed this large loop of ionised gas (prominence) from the Sun's outer atmosphere (corona) with a smaller 4in. (10.16cm) telescope.

By comparing photographs of the stars near the Hyades several months before, during and after the eclipse, the Greenwich teams could detect a tiny average deflection of 1.64 arcseconds – just a fraction on the plates – which confirmed Einstein's theory that the gravitational mass of the Sun would distort nearby starlight twice as much as predicted by Newtonian physics.

After months of calculations, Dyson and Eddington presented their results to a packed audience of scientists on 6 November 1919, leading *The Times* to proclaim the next morning, 'REVOLUTION IN SCIENCE'. For Eddington, it was not just a scientific success but also one of Anglo-German collaboration in the aftermath of war. He finally met with Einstein in London during June 1921.

## MEASURING DEVICE FOR THE STAR TRAIL CAMERA PLATES

Standing 14in, (35cm) tall when fully opened, this measuring device for photographic glass plates is a rare surviving relic of a special type of telescope that astronomers used both to select and close observatory sites, including Greenwich. The pole star trail telescope was originally developed in the 1880s by astronomers at Harvard College Observatory as a means of assessing sky conditions (referred to as 'seeing') when measuring the absolute brightness of stars (photometry). Another version was later used by meteorologists at Blue Hill Observatory near Boston and the US Weather Bureau in Chicago for measuring cloud cover. It was for this second purpose that such an instrument was installed at Greenwich in 1920.

Known as the Night Sky Recorder, the combined telescope and camera assembly was situated on Bradley's meridian (17) and directly aligned with Polaris. Depending on your latitude, certain stars around Polaris are circumpolar and never set below the horizon. At sunset each night the watchman removed the telescope's cover and replaced it shortly before dawn. A few hours later, one of the Meteorological Assistants would develop the long-exposure photographic plate before installing it within the central holder of this device. The image revealed circular trails of stars around Polaris, and light reflected from the mirrored base was used to measure the hourly duration of each trail on the 24-hour scale. Clear skies produced long, well-defined arcs whereas cloudy nights produced trails that were either intermittent or completely absent. The Assistant then sent the results to the Met Office by telephone.

By the 1930s, the meteorological instrument readings were starting to confirm the astronomers' own experience of deteriorating observing conditions at Greenwich and provided compelling evidence in support of the tenth Astronomer Royal's recommendation to relocate to Herstmonceux (see *Epilogue*). In his Annual Report for 1953, Spencer Jones used the data from the Night Sky Recorder at the new site to conclude that 'the number of hours of clear night sky was about ten per cent higher at Herstmonceux than at Greenwich'. In a curious twist of fate, another type of pole star trail telescope – designed to measure atmospheric turbulence rather than cloud cover – was used to select a new observatory site on La Palma in the Canary Islands, which ultimately led to the closure of Herstmonceux itself in 1990.

Measuring device for the star trail camera plates with a detailed view of the scale, unknown maker, about 1920 (AST0736)

## REGULATOR No.2016 BY DENT

This regulator (see *How does a mechanical clock work?*) is a classic example of how Observatory instruments were modified and repurposed over several decades to meet changing scientific and national requirements. It was originally commissioned as the reserve precision timekeeper for Station 'D' in New Zealand for the transit of Venus expedition in 1874 (see *Transit of Venus expeditions*). Two decades later, in April 1892, Chief Assistant Herbert Turner transported this clock to Waterville on the west coast of Ireland where the regulator was connected to the transatlantic submarine cable network (see *Cable production in Greenwich*) to help astronomers send time signals to North American observatories for comparing longitudes. Upon its return to Greenwich, the timekeeper was shuffled between locations until 1897 when it was housed alongside the Thompson Equatorial telescopes (82) atop the New Physical Observatory (78).

In 1924, 'Dent 2016' gained new significance when it was chosen as the timekeeper for the BBC radio time signals known today as 'the six pips'. It was an opportune moment for a new public time signal as the Summer Time Act of 1922 continued the First World War experiment of changing the clocks (daylight saving) to reduce coal consumption. Based on a suggestion by the horologist Frank Hope-Jones (93), the BBC's General Director, John Reith, contacted the Astronomer Royal about improving the accuracy of radio time signals to help the public adjust its clocks. The astronomers and technicians at Greenwich decided to send time signals via telephone to the BBC's Savoy Hill studios every 30 minutes. Originally known as the 'six dots', the now familiar tones were broadcast twice a day to count down the last five seconds of the hour, with the final tone on the hour itself. Equipped with electrical contacts from its previous functions, Dent 2016 was an obvious choice for the task and the first radio transmission was made at 9.30 p.m. on 5 February 1924, introduced by Dyson himself.

Listeners appreciated this innovative service and the regulator continued to send the signal to the BBC for another 25 years, despite being relocated for safety to Abinger in Surrey (91) during the disruptive years of the Second World War.

An eight-day wall-mounted astronomical regulator no.2016, Edward John Dent and Co., 1874 (ZAA1293)

## SHORTT FREE-PENDULUM CLOCK SYSTEM

On 1 January 1925, a different type of clock came into use as the Observatory's new standard timekeeper for sidereal time. Railway engineer William Shortt first became interested in precision horology after designing an electrical recorder for timing the speed of trains. A chance encounter with Frank Hope-Jones of the Synchronome Company (93) in 1910 inspired him to design an electromechanical timekeeper that subsequently became the most accurate pendulum clock commercially available.

The system was composed of two parts: the 'master' pendulum was set within a partially evacuated copper tank, topped by an electric impulse mechanism and then sealed within an airtight glass cover. By isolating the moving parts from the effects of air resistance and changes in atmospheric pressure, Shortt could ensure a consistent pendulum swing of one second. The engineer also chose to construct the master pendulum from Invar, a nickel-iron alloy that was unaffected by temperature variation. Every 30 seconds, the electric 'slave' regulator energised the master clock by transmitting an impulse at the lowest point in the pendulum's swing for minimal disturbance. In return, electric impulses from the master clock adjusted the slave clock's pendulum to keep accurate time.

The first Shortt system was installed at the Royal Observatory Edinburgh in December 1921. It performed exceptionally well, losing just one second over the course of a year and prompted Dyson to order one for Greenwich. The master clock of 'Shortt 3' was installed in the cellars of Flamsteed House, where the thick walls provided stability and a constant ambient temperature of 17.1°C (62.8°F) while the slave clock was installed in the Meridian Observatory for ease of access.

Over the next 20 years, around 100 Shortt systems would be installed at observatories around the globe, including a further five at Greenwich. As astronomers began to inspect several years of Shortt data, they realised that the Earth's rotation was not as constant as previously measured (3), confirming long-held suspicions. They combined readings from multiple Shortt systems to create a mean average sidereal time but the 250-year dominance of precision pendulum technology was coming to a close. Scientists were now turning their attention to laboratories to devise new, more accurate time standards based on quartz (99) and, later, atomic technologies (100, see *Epilogue*).

'Shortt no.3' free-pendulum clock system with glass cover removed, Shortt and the Synchronome Company, 1924 (ZAA0539)

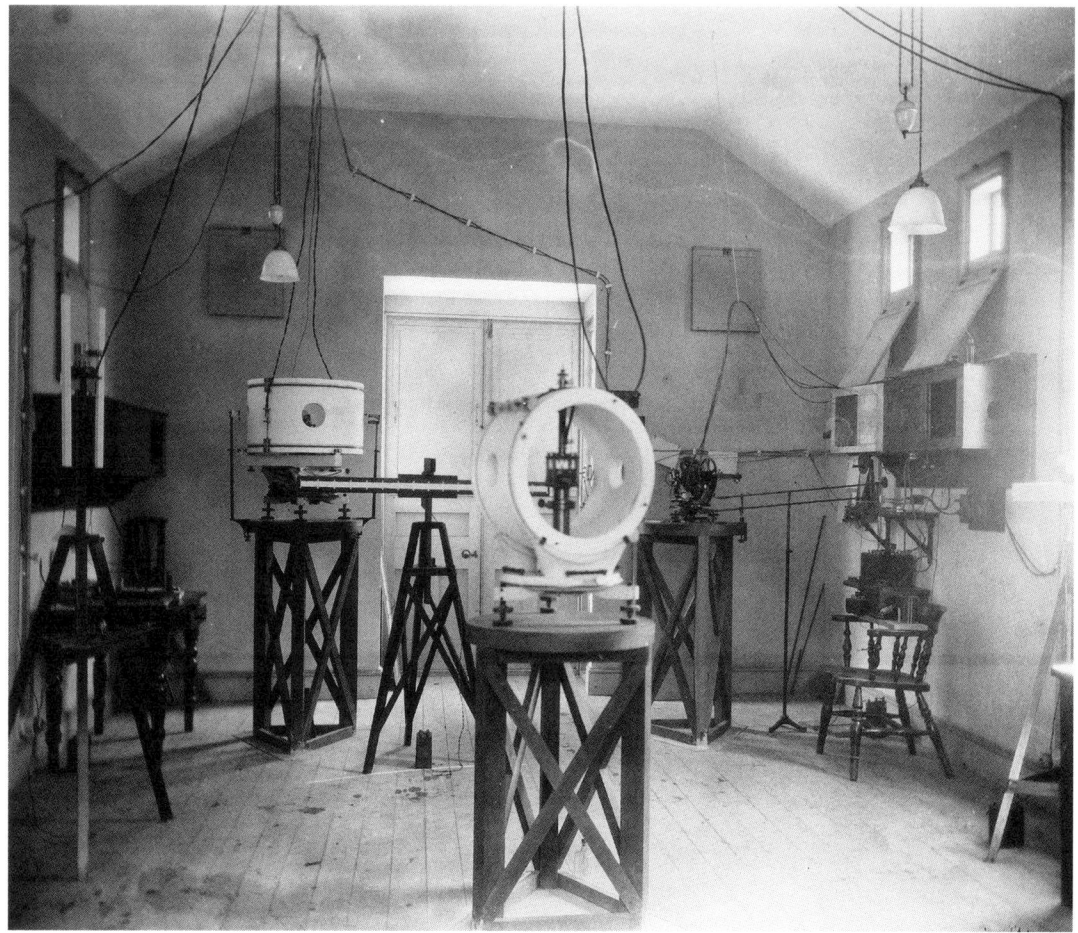

91

GLASS PHOTOPOSITIVE OF THE MAGNETIC PAVILION AT ABINGER

As the Observatory faced increasing disturbance from the vibrations and electrical interference of nearby railway lines, Dyson started to investigate the relocation of the Magnetic Observatory. Working with the Admiralty and the Board of Visitors, he negotiated some compensation from the local railway company and used the funds to establish a new site at Abinger in Surrey, about 30 miles (48km) south-west of Greenwich.

The buildings were constructed of carefully chosen non-magnetic materials, with the instruments supported on concrete piers to insulate them from the floor. This photograph, believed to have been taken around the time of the first observations in March 1925, shows the layout of the instruments in the main Magnetic Pavilion.

The two drum-shaped instruments, known as coil magnetometers, were used by the Abinger scientists to measure variations in the Earth's magnetic field. One instrument (centre) detected horizontal variations, while the second (back left) responded to vertical variations. Another small instrument (back right), a dip circle (54), was used to

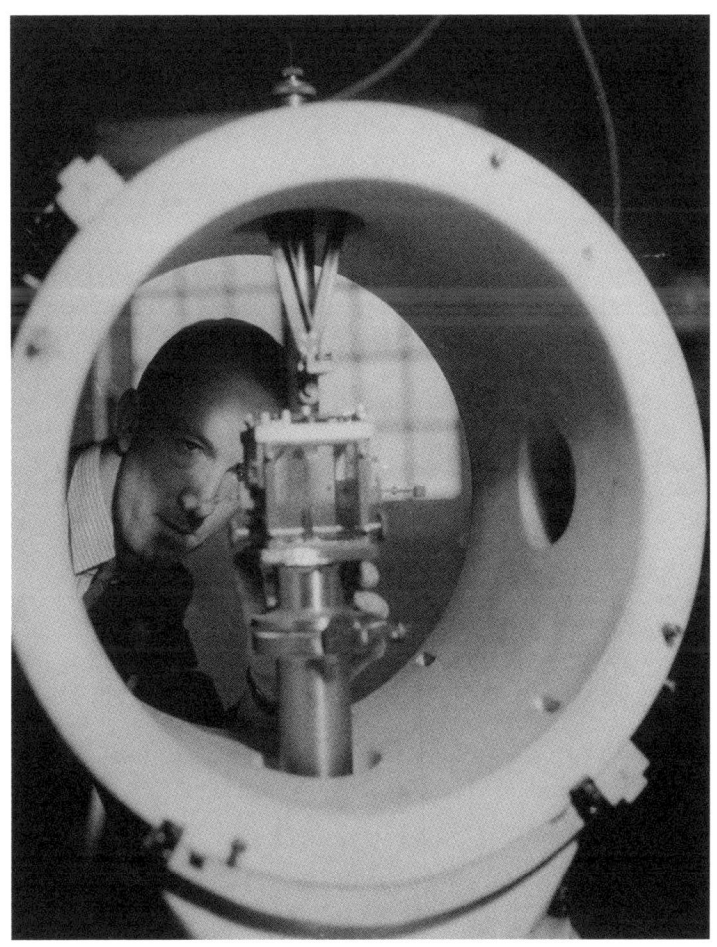

⊚
The newly constructed Magnetic Pavilion at Abinger, Surrey, William Moody Witchell, about 1925 (AST1119)

⊚
P. Rickerby examines a coil magnetometer, unknown photographer, about 1945 (REG18/000454.42)

measure the local horizontal angle between a magnetised needle and the Earth's magnetic field lines.

The observers used a series of lamps and mirrors to view the tiny movement of the magnets suspended within each drum against graduated scales on the piers behind. A theodolite was mounted on another pier so that they could observe the deflections from a distance without disturbing the sensitive instruments. The nearby Magnetograph House contained a series of clockwork-powered drums covered in light-sensitive paper that moved at a steady rate to automatically record the reflected light from the instrument scales. A series of shuttered openings were used by the staff to allow them to wind the clocks and to replace the photographic paper with minimal disturbance.

Initially, the new facility was a success, with a growing team of staff members. But the problems that had endangered observations at Greenwich soon started to affect the results at Abinger, particularly after the electrification of the Mole Valley Railway in 1938. The Magnetic Observatory moved once again in 1957 to Hartland, north Devon, where it continues to operate under the British Geological Survey.

## OBSERVER'S CARD FOR THE GIGGLESWICK TOTAL SOLAR ECLIPSE

For Observatory Assistant Philibert J. Melotte (82), this yellow card was his most precious possession on 29 June 1927 as he pushed through a 100,000-strong crowd assembled in Giggleswick, Yorkshire. This was the first total eclipse to sweep across Britain in two centuries and its fleeting 23 seconds of totality – when the Moon completely obscures the Sun – would only be visible in a band 30 miles (48km) wide that stretched from North Wales to Hartlepool in County Durham.

Protected by Army cadets, the Greenwich observers – Melotte, Jackson and Davidson – checked their equipment for the last time, with Observatory Foreman Woodman on hand for any last-minute repairs. The main instrument was a telescope with a 6in. (15.2cm) lens and 45ft (13.7m) focal length that was mounted horizontally in a metal lattice and covered with protective dark cloth and waterproof canvas. The image of the solar disc was reflected into the stationary telescope via an 8in. (20cm) mirror that tracked the Sun (a coelostat). Two spectroscopes (see *Splitting starlight with spectroscopy*) were used to measure certain chemicals within the Sun's outer atmosphere (or corona).

Despite a gloomy start, the clouds began to disperse at 5.50 a.m. with just 10 minutes to spare. The countdown continued with a chronometer checked against the BBC's six pips (89) and then, as totality began, Davidson called out 'Go', to signal the start of the photographic exposures. Dyson watched from the side, accompanied by his wife and daughter Ruth, who commented on the unsettled behaviour of the nearby birds and sheep. A few hours later, Melotte had successfully developed Davidson's photograph of the solar corona as seen during totality and the image was sent by aeroplane to London where it was published in the *Evening News* later that day.

Dyson's other observers were less successful. Assistants Witchell and Greaves viewed the eclipse from above the clouds, courtesy of an Imperial Airways plane chartered by the *Daily Mail*. Flying over their colleagues at 11,700ft (3,566m), they used two cameras, one mounted on the roof and another through an open window. Despite the addition of shock-absorbing material, the cameras shook and the images blurred. Similarly, Assistants Edney and Chamberlain went aboard HMS *Fitzroy* to view totality from the North Sea but were clouded out and returned with nothing.

Eclipse LNER railway poster, unknown artist, 1927
Science and Society Picture Library

John Jackson with the telescope at Giggleswick, A. Horner and Sons, 1927
(AST1096)

Solar eclipse official observer card, unknown maker, 1927 (AST0050.4)

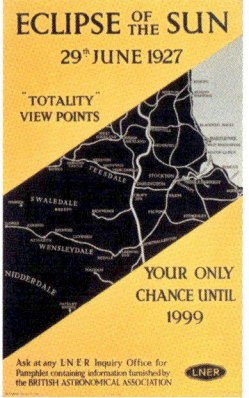

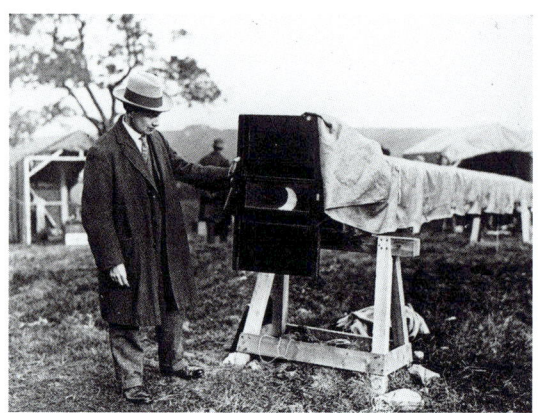

# SOLAR ECLIPSE.

## 29th JUNE, 1927.

## Official Observer.

F. W. DYSON,
Astronomer Royal.

W. HUDDLESTONE,
Superintendent.

To be shown if held up by Police.

## 93

## WARREN SYNCLOCK

Plugged into the mains and whirring away silently on the mantelpiece, the electric 'Synclock' provided domestic users with the technology to keep accurate time at home. After the First World War only six per cent of British homes had electricity, but by the late 1930s it was almost 70 per cent. The London-based firm of electrical engineers Everett Edgcumbe used the patent by American inventor Henry Warren to create a domestic clock that could be regulated by mains electricity. Once set to time manually, the clock continued to run with no batteries or winding, making it a popular choice for many.

The Synclock's self-starting motor revolved over 4.3 million times a day, perfectly in sync with the power stations that relied on GMT radio signals from Rugby to regulate a consistent supply of electric current at 50Hz (50 cycles per second). These radio time signals were directly sourced from the Observatory's Shortt Clocks (90) via a telephone link and were broadcast twice a day at 10 a.m. and 6 p.m. Rugby was home to the world's most powerful radio transmitter and its long-wave 16kHz radio signals, known as the

Ethel Cain, first voice of the speaking clock, unknown photographer and date
BT Archives

Warren Synclock, Everett Edgcumbe Company Ltd, about 1931 (ZBA1774)

International Time Service (call sign 'GBR'), provided a reassuring and highly accurate measure of GMT to navigators across the globe from December 1927.

Another option for checking accurate time at home was to ring the 'Speaking Clock' to hear the recorded tones of London telephonist Ethel (Jane) Cain, whose 'golden voice' was selected from over 15,000 applicants. Hosted by the General Post Office (GPO), the automatic service began on 24 July 1936 and was instantly popular, with over 20 million calls in the first year generating much income and conveniently saving GPO operators from bothersome enquirers requesting the time. The Speaking Clock initially relied on hourly synchronisation with time signals sent directly from the Observatory before switching to the Rugby radio signals in 1963. Today, the Speaking Clock is provided by phone networks and is regulated by atomic time signals from the National Physical Laboratory. It is particularly popular on four key dates each year: New Year's Eve, the days when the clocks go forwards and backwards, and for timing the two-minute silence on Armistice Day.

## RUPERT GOULD'S NOTEBOOKS

While astronomers at Greenwich were busy adapting to new timekeeping technologies in the 1920s, other experts were more concerned about the decay of the site's historic timekeepers. On 5 March 1920, Lieutenant Commander Rupert Gould came to see the marine timekeepers by John Harrison as research for his forthcoming publication, *The Marine Chronometer* (1923). He was dismayed to find them 'dirty, defective and corroded' and persuaded the Astronomer Royal to let him restore the first one, 'H1' (13), at his home workshop in Kensington. Gould was no professional horologist but he had first-hand experience of using chronometers at sea and was a fellow member of the British Horological Institute, hence Dyson's agreement to the request. At that time, Gould was employed by the Hydrographic Office and could only work on the Harrison timekeepers in his own time. Much later, this generous offer would become a burden as Gould's life unravelled with ill-health, the breakdown of his marriage and loss of his home and job.

Having safely cleaned and returned the instrument by Christmas 1920, Gould felt emboldened to seek authorisation to work on 'H4' (21) and so began his ambitious and somewhat obsessive 13-year project to restore all four marine timekeepers to full working order. His early work on 'H1' and 'H4' was undocumented but for his later restoration of 'H2' (14) and 'H3' (18), he undertook a more structured approach, carefully documenting his progress across a series of letters and notebooks. Filled with detailed annotations and carefully executed technical drawings, they provide us with an insight into Gould's mercurial mind as he scribbled, calculated and crafted replacement parts. Although Gould's choices and methodologies may not always have chimed with either contemporary or modern conservation techniques, he undoubtedly brought the Harrison timekeepers back to life, enabling us to enjoy their astonishing story and mesmerising motion today.

➜ Rupert Gould, unknown photographer and date
Sarah Allan

➜➜ Notebooks on the reconstruction of Harrison's first three marine timekeepers, Rupert Gould, 1923–33
(GOU/1, GOU/4)

| Wheel tested. | | Remarks |
|---|---|---|
| 1 | | Free, but could do with a little broaching, thickest oil. |
| 2 | | Free. Polished pin. |
| 3 | | Free. Polished pin. |
| 4 | | Free, but gave a touch with the broach, & polished |
| 5 | Fitted safety | Touched up with broach. Arbor straightened & polished. |
| 6 | ratchet to prevent unmeshing | Slightly straightened & polished arbor. Slightly broached L.H. arbor. |
| 7 | | Free. Polished pin. |
| 8 | | Free. Polished arbor pin. |
| 9 | | Free. Polished arbor pin. |
| 10 | | Free. Polished arbor pin. |
| 11 | | Free. Slightly straightened arbor & polished it & pin. |
| 12 | | Broached slightly. Polished arbor pin. |
| 13 | NEW | Free, but broached L.H for safety, this being new. Polished pin. |
| 14 | | Free. Polished arbor pin. |
| 15 | NEW. | Free, but broached, being a new wheel. Polished pin. |
| 16 | | Free. Polished pin. |
| 17 | Fitted safety washer to 17 to prevent it touching 18 | Not too free. Broached arbor & filed ends. Reversed & repinned. |
| 18 | | Free. Polished Pin. |
| 19 | | Free. Polished pin. |
| 20 | NEW | Definitely stiff. Broached arbor & polished pin. |
| 21 | Fitted safety washer to 21 to prevent touching 19 | Free. Polished pin. |
| 22 | | Free. Polished pin. This examination finished, 6·20 pm 25·VIII·34 |
| 23 | NEW | } Heavily broached. Polished arbors, & fitted safety washer to 24. |
| 24 | NEW | |

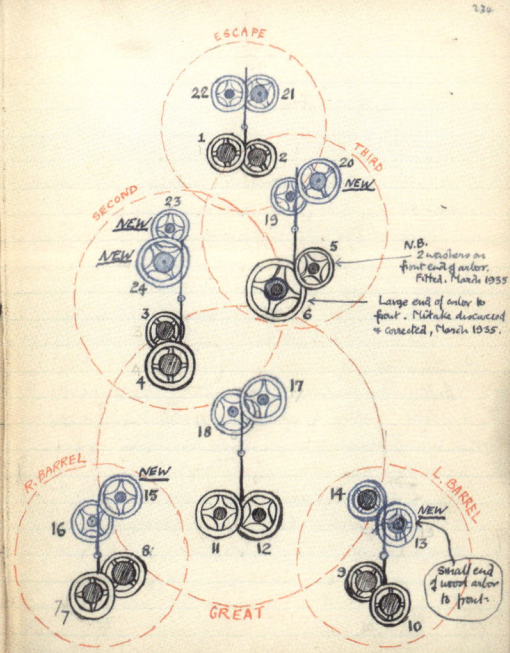

---

56

A under B, marked on rim. (A A)
B over A,  "   "  rim.
C under B, "    "  rim. (C C)
D over C,  "    "  hub.

E & G under F, all marked on rim

H under I, H on rim, I on hub.

J under K (both under H & I).
J on rim, K on hub.

L under M, L on rim, M on hub.

N & O on rim. N under O.

Pillared plates marked with reference letter.

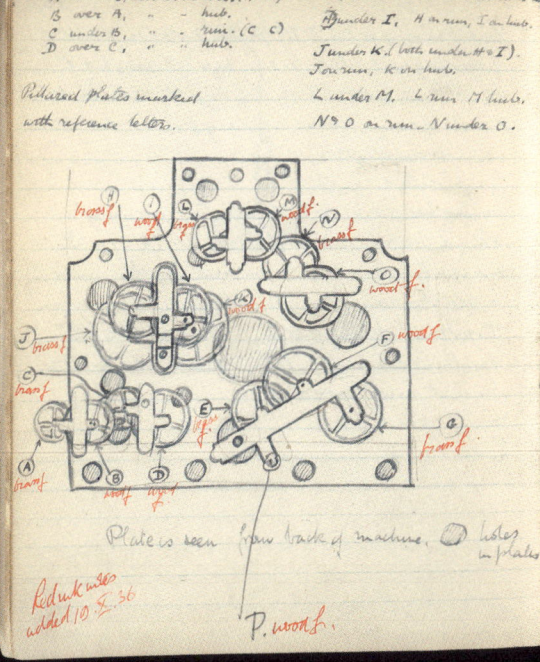

Plate is seen from back of machine. ⓞ holes in plates

Red ink rules added 10·X·36

P. wood f.

---

57

13·1·24.  I tested the torque of the remontoire springs today, and found they would only support a wt of ·8 oz, half what is now req'd to keep machine going. Well, I must do what I can to bring the driving torque down still further.

I shall not be able to work on the machine proper for some time, as I have to do 33,000 odd words of Spanish translation for Davy. I may clean away some parts for cleaning meanwhile.

10·2·24. Owing to an opportune breakdown of my type-writer, I have a week-end, or rather Sunday, in which I can do a little. Shall start cleaning the plates.

Removed friction wheels of 3rd plate, marking them as shown opposite.

Mary French, unknown photographer, 1933
(AST1113.31)

## MARY FRENCH WITH THE PLATE MICROMETER

Taken in the months leading up to Dyson's retirement, this photograph embodies the changing profile of the Observatory's staff during the early twentieth century. Shown here at her desk in the New Physical Observatory (78), French was one of three additional women recruited in 1930 to work on the photographic plates generated by the Thompson Equatorial telescopes (82) and the Astrographic telescope (71). By comparing the star positions on each plate with those recorded 30 years previously as part of the Carte du Ciel project (71), astronomers could measure the long-term shift in star positions. By looking at each plate through the microscope and adjusting a series of wires, French and her colleagues could each measure the coordinates of several stars per hour using this table-top machine.

Apart from the five-year period of Christie's 'lady computers' in the 1890s (71), women had no opportunity to work at the Observatory until the First World War. After the departure of the majority of the (male) computers for military service (84), the Admiralty conceded that Dyson could recruit women to fill some of the vacancies. By March 1917, only two women – E.D. Lang and E.W. Clack – had been appointed, leading Dyson to enlist his daughter Margie to undertake the nightly magnetic and meteorological readings (although she was seemingly scatty and had to rely on a notice pinned to her bedroom wall: 'Remember Meteorology'). After the war, Andrina Crommelin, daughter of Assistant Andrew Crommelin (77), joined the Photoheliograph Department (64), where she took daily photographs of the Sun and measured sunspots until 1925.

By the outbreak of war again in 1939, there were 14 women working across various Observatory departments. Most female staff only stayed for around three to five years with some exceptions, most notably Clack, who continued with the daily 'mag and met' readings for 20 years. French continued working on the Thompson plates until her marriage in 1939. Unaffected by the restrictions faced by her predecessors in the 1890s (73), she returned as 'Mrs Higby' in 1941. She briefly worked alongside Philip Laurie (98) in the Photoheliographic Department but, sadly, poor health forced her to resign again just a year later.

# Harold Spencer Jones
## 1890–1960

Born in Kensington, London, Harold Spencer Jones was actively encouraged to study mathematics by his parents and later excelled in his studies at Jesus College, Cambridge. He joined the Royal Observatory in July 1913 as Chief Assistant and began studying stellar magnitudes. His work was soon interrupted by the First World War, when, like many other observatory colleagues, he was reassigned to work on optical instrument design for the Ministry of Munitions. A few years later, Spencer Jones married fellow Londoner Gladys Mary Owers and the couple moved to South Africa in 1923 upon his appointment as Astronomer at the Cape Observatory.

A decade later, Spencer Jones became the tenth Astronomer Royal and the family returned to Greenwich. His first task was to oversee the installation of two new large-scale telescopes for better astrophotography and improved time standards. As observing conditions began to deteriorate, he recommended the relocation of the Royal Observatory but his plans were delayed until after the Second World War. He subsequently took the lead in choosing Herstmonceux in Sussex for the Observatory's new home and organised the move (see *Epilogue*).

Beyond his immense administrative and organisational work, Spencer Jones was renowned for several major contributions to science. While working at the Cape Observatory, he coordinated a global network of astronomers to observe the close approach of the asteroid Eros in 1930–31. Over the next decade, he collected the observations and reduced the data from multiple observatories to improve the value of the distance between the Earth and Sun (the astronomical unit or au). He also used thousands of observations of the Sun, Moon and planets to detect irregularities in the Earth's rotation on its axis, the results of which contributed to the development of Ephemeris Time (ET) in the 1950s.

More generally, Spencer Jones was involved in various community projects such as improving practical training for the watch industry and using geomagnetic research to aid navigation. After his retirement in 1955, he continued to advocate for greater international collaboration in science and died at his Kensington home in November 1960.

Harold Spencer Jones, Elliott and Fry Ltd, about 1944 (ZBA1819)

## SLITLESS SPECTROGRAPH

Just one year after his appointment, Spencer Jones found himself in the public spotlight as the Observatory's newest equatorial telescope came into service. On Visitation Day, 2 June 1934, the First Lord of the Admiralty opened the new Yapp Dome and Telescope, situated a short distance east of Flamsteed House in Greenwich Park. Made by Grubb Parsons, the 36in. (91cm) reflecting telescope was a gift from the cigarette manufacturer William Johnston Yapp in recognition of Dyson's retirement as ninth Astronomer Royal. Unable to keep up with their counterparts at American observatories who were installing immense instruments under clear Californian skies, the astronomers at Greenwich also decided to use Yapp's generosity to commission a spectrograph (see *Splitting starlight with spectroscopy*) that could extend their astrophysics research on the colour temperatures of stars. According to the classification system developed in 1913 by Ejnar Hertzsprung and Henry Norris Russell, blue-white stars are hot, young and luminous whereas red ones are cooler, older and less luminous. Our Sun is a middle-aged (main sequence) star that predominantly emits yellow light.

At 5ft (1.5m) tall, this slitless spectrograph, designed to enable analysis of multiple stars within the field of view, was just one small component of the overall telescope assembly that extended over 25ft (9m) high. Once attached, a 6in. (15.2cm) parallel beam of light from the telescope was directed through the spectrograph via a 45° prism and 9in. (22.9cm) concave mirror before being focused onto a photographic plate at the side. An electric gramophone motor and a series of rotating discs enabled the astronomers to disperse each star's spectrum for ease of analysis, while a black fabric cover inside the lid excluded any stray light. A carbon arc lamp on the roof of Flamsteed House, calibrated to emit light at a specific wavelength, provided a terrestrial comparison against the stellar spectra.

In November 1938, after a few years' service, the slitless spectrograph was replaced with a slit version that could be used more selectively on certain stars of interest at ultraviolet wavelengths. The astronomers successfully measured over 250 stars but the worsening atmospheric conditions at Greenwich, coupled with wartime searchlights, compelled them to stop in 1939.

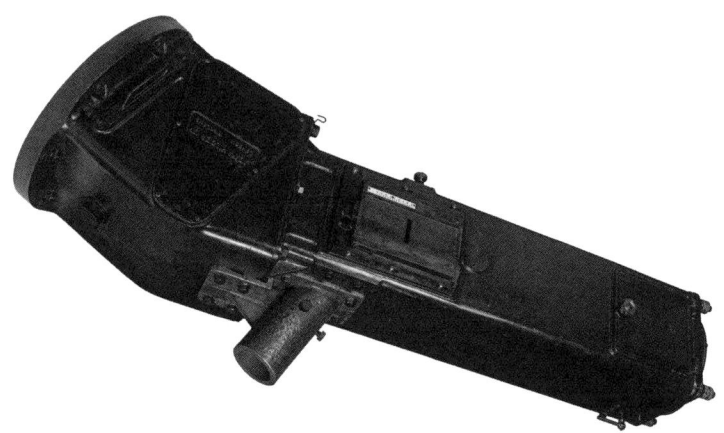

◗
Slitless spectrograph for the Yapp Telescope, Adam Hilger Ltd, 1934 (AST1177)

◗◗
Postcard showing the Yapp Telescope with the slitless spectrograph attached, unknown maker, 1934 (AST1107.25)

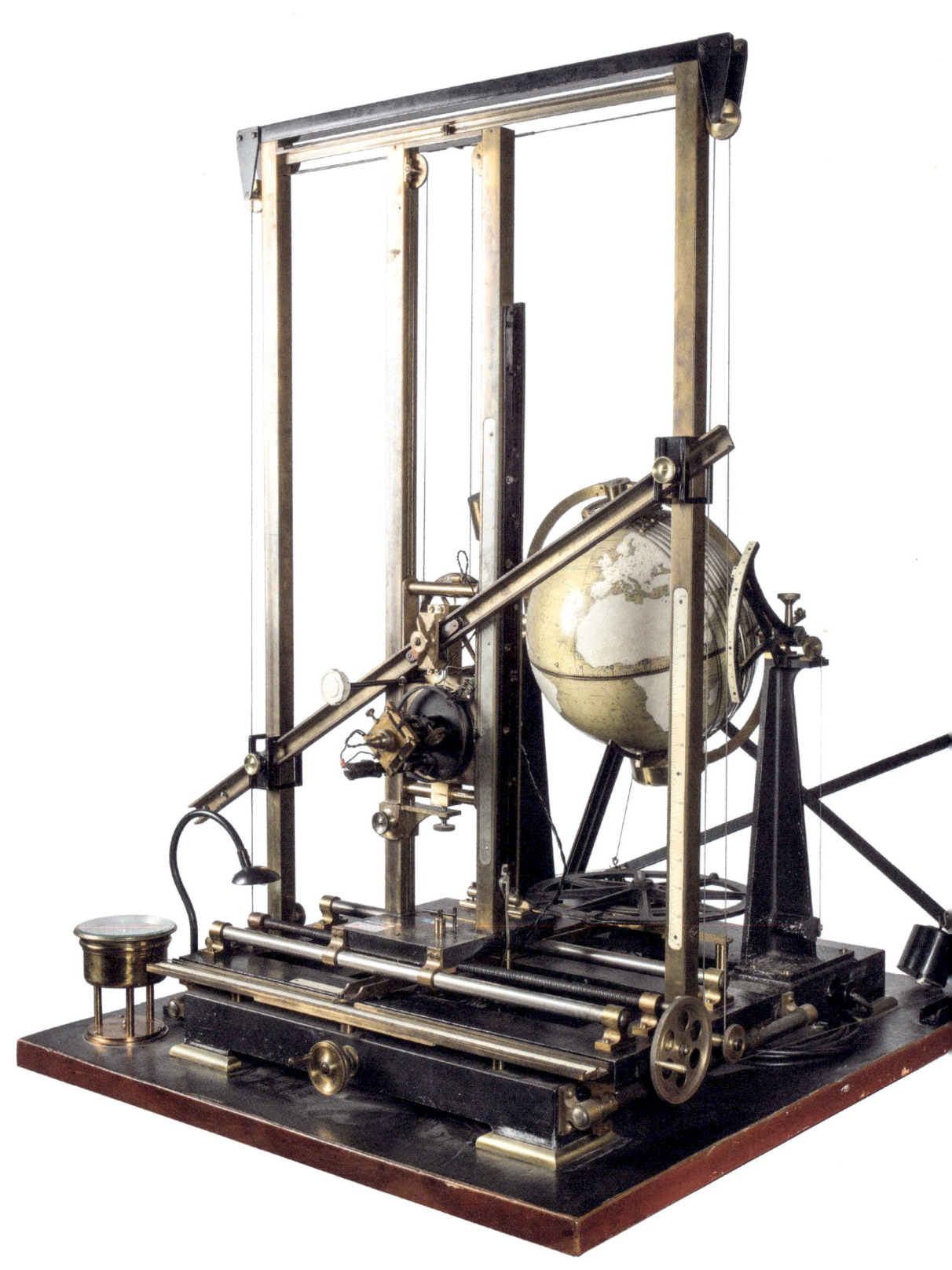

## 97
## OCCULTATION MACHINE

This quirky and unique object beautifully illustrates the mid-twentieth-century transition from analogue to electronic computing at Greenwich and observatories elsewhere. By the 1920s, both amateur and professional astronomers were collecting observations of lunar occultations – when the Moon passes in front of a bright star for a few minutes – to help refine our understanding of the Moon's complicated orbit. One such amateur was James Duncan McNeile, who in 1929 created an analogue computer composed of a globe with a wooden frame that could be used to predict when and where such occultations would occur. He shared the data with fellow members of the British Astronomical Association (BAA, 81), one of whom was John Comrie, Superintendent of HM Nautical Almanac Office (HMNAO, the modern continuation of the Observatory's navigational work from the 1760s). Comrie had already revolutionised HMNAO's work by adopting commercial calculating machines and punched card equipment, and so he combined his amateur and professional interests to persuade his superiors to fund the construction of a more robust occultation machine made of metal. The Admiralty approved and the task was assigned to A.C. Wescott, Chief Joiner of the Observatory's workshop.

The machine, 6ft (1.8m) tall, was situated in darkness behind a curtain. The operator set the 12in. (30cm) diameter globe to the angle of the occulted star before adjusting the diagonal crossbar to the angle of the Moon's orbit. Light from a small car headlamp was then directed through a 4in. (10cm) condenser lens to create a disc of light on the globe's surface. As the operator slowly rotated the globe, he used the hour pointer to predict the timings for each of the 60 observing stations up to three years in advance, and called out the results to a colleague sitting on the other side of the curtain.

The machine was a remarkable success: it took just two minutes to set up and provided predictions, accurate to within a minute, for all stations. With a few modifications, the machine remained in use until it was replaced by ICT 1201 electronic computers in the mid-1960s.

Occultation machine and detail of Europe as seen on the globe, A.C.S. Wescott, 1934 (NAV1803)

## PHILIP LAURIE'S WARTIME DIARIES

On 14 May 1940, radio listeners heard Anthony Eden, Secretary of State for War, announce the creation of the Local Defence Volunteers, which became known as the Home Guard. With permission from the Admiralty, 18 members of Observatory staff created their own unit, headed by the Astronomer Royal himself. One of the volunteers was Junior Assistant Philip Laurie, who decided to record the unit's activities in two notebooks, which offer a fascinating eyewitness account of wartime life at the Observatory.

Laurie's choice of title for his first volume – 'The Beleaguered Garrison' – is typical of his witty and irreverent journalistic style as he documented his experience with photographs, cuttings and ink sketches. At first, the unit focused on patrol schedules, uniforms and practicing with rifles at the Home Guard's training range at Bisley, Surrey. But the drills became real with the onset of the Blitz in September 1940, which Laurie described as 'leaving fires everywhere and black smoke pouring'. By the end of the month, the unit had been depleted with the evacuation of the Astronomer Royal and most of the scientific staff to Abinger in Surrey (91), leaving Laurie and others to join forces with the Greenwich Park Home Guard.

On 15 October 1940, a bomb destroyed the Observatory's main gates, damaging both the Shepherd Gate Clock (49) and nearby Equatorial dome (75) and leaving a small crater that added to the Park's increasingly pockmarked appearance. Apart from some additional bomb damage to the Altazimuth Pavilion (77), the Observatory survived the war relatively unscathed. The Home Guard was disbanded in December 1944, thus depriving Laurie of his small allowance that was spent with 'entirely patriotic motives' in the nearby pub, the Rose and Crown.

Laurie's second volume – 'Scientia Pro Patria' – documents the initial wartime experience of the Observatory employees as they filled sandbags, removed telescope components for safekeeping and gained essential skills in First Aid and firefighting. After the evacuation, however, Laurie's comments became increasingly sarcastic as he described the despatch of fruit and vegetables from the Observatory's gardens to Abinger and the complaints of the evacuated staff whose sleep was disrupted by German bombers flying overhead. As one of the Observatory's remaining defenders, Laurie had good reason to offer little sympathy.

Pages from wartime diaries by Philip Laurie; the group photo shows (from left to right) Laurie, Norman Rhodes and George Wells, unknown photographer, 1939–45
Cambridge University Library

## III PARK SECTION

Meanwhile the Greenwich Park Section had been formed to cover the Park, being drawn from Keepers, gardeners, etc, employed (the Section leader was Austin, the Superintendent) and totalled 28 men.

Austin now proposed that the remnants of the R.O. section should join the Park Section and Rhodes a comparative new-comer to the R.O. joined the Unit. Shell Wells and Lousie became attached somewhat later, in December. The last pair had kept up some sort of guard at the R.O. after Sept 25th but the situation was quite impracticable and they accordingly fell in with Austin's plans.

Nestor refused to transfer to the Park Section on the grounds that his authority as liaison between the A.R. and the Home Guard would be impaired and that it would interfere with administrative duties.

Two rooms were allocated to the Park Section (the Assistants Room to be used as Guard Room and, reluctantly, the upper room for emergencies) and they maintained three men on guard nightly who agreed to keep an eye on the Octagon Room in exchange for the 'concession'.

On Oct 15th at 10.20 p.m. an HE bomb destroyed the Main Gate & 24-hour dial, cutting the services, and reducing all glass work to fragments. The gas leak was repaired by Mountain (acting foreman) and a watch was kept by ARP & HG over the hole with due solemnity. Mountain was awarded the B.E.M. for the night's work. The second dose of HE occurred on Oct 21st when about 18 small HEs fell across the R.O., damaging

## 99
## QUARTZ CLOCK AND FREQUENCY STANDARD

This nearly 6ft (1.8m) tall rack of wires, flashing valves and moving dials is the most unlikely looking clock in the Museum's collection, yet it represents a fundamental shift in timekeeping during the 1930s. While observatories around the globe were busy installing Shortt free-pendulum systems (90), scientists in laboratories in Britain and the US were taking inspiration from radio technology to explore how applying an electric current to a piece of quartz (silicon dioxide) could induce it to vibrate at very accurate and stable frequencies.

In his Annual Report for 1937, Astronomer Royal Spencer Jones raised concerns about the weather: during cloudy periods, there was no opportunity to observe the stars to check the Shortts, leading to an error of 0.05 seconds per day, compared to the 0.001 seconds per day typically lost by quartz oscillator systems. He was keen to invest in these highly stable timekeepers that required fewer observations and managed to secure funding for the Observatory's first quartz clock, 'Q3', which was made by the National Physical Laboratory and installed at Greenwich in early 1939.

Within two years, however, the Time Department was relocated to Abinger in Surrey (91) to sustain its essential national service away from the Blitz attacks on London. The quartz clock 'Q3' was transferred to Abinger in 1943 and staff there began to combine data from various Shortt and quartz clocks nationwide to create an extremely accurate mean average clock. Scientists were beginning to realise that the age of the precision pendulum clock was coming to an end.

The quartz clock shown here was designed by the Post Office Research Station in 1944 and was used for regulating time signals transmitted from Rugby (93) before its conversion into a teaching model for technical colleges. Vibrating at 100 kilocycles per second, the crystal in the cylinder at the top of the rack sends a signal to a series of circuits, amplifiers and motors that can divide or multiply the signal into frequencies suitable for controlling clocks. The quartz movement in your wristwatch – busy oscillating 32,768 times a second – is a descendant of this technology.

◉ (and details)
Quartz clock and frequency standard, designed by General Post Office Engineering Research Station, manufactured by Roberts & Armstrong, about 1944
(ZAA0289)

Replica of HP 5061A caesium-beam frequency standard atomic clock, Hewlett-Packard, 1973 (ZAA0549)

## 100
## HP5061A CAESIUM-BEAM ATOMIC CLOCK

As observatories worldwide made the transition from pendulum to quartz clocks (99) in the 1940s, an even bigger revolution in timekeeping was already underway. At the National Physical Laboratory (NPL) in Teddington, London, Louis Essen was using his wartime work on microwaves and light to develop an atomic clock, spurred on by similar developments in the US. Essen and his colleague John Parry eventually created a timekeeper in which caesium-133 atoms were fired into a vacuum chamber to be energised by a microwave beam, the frequency of which was adjusted to match the atoms' natural frequency, causing them to resonate and emit light. A detector used these light pulses to regulate a quartz oscillator to a steady rate. Essen and Parry powered up the world's first operational caesium atomic clock at the NPL on 24 May 1955 and measured one second with unparalleled accuracy as 9,192,631,830 energy transitions.

Within four years, the Observatory's Time Department – now relocated to Herstmonceux (see *Epilogue*) – was reliant on daily updates from the caesium clocks at the NPL to regulate its quartz clocks. At the same time, engineers at Hewlett-Packard (HP) were

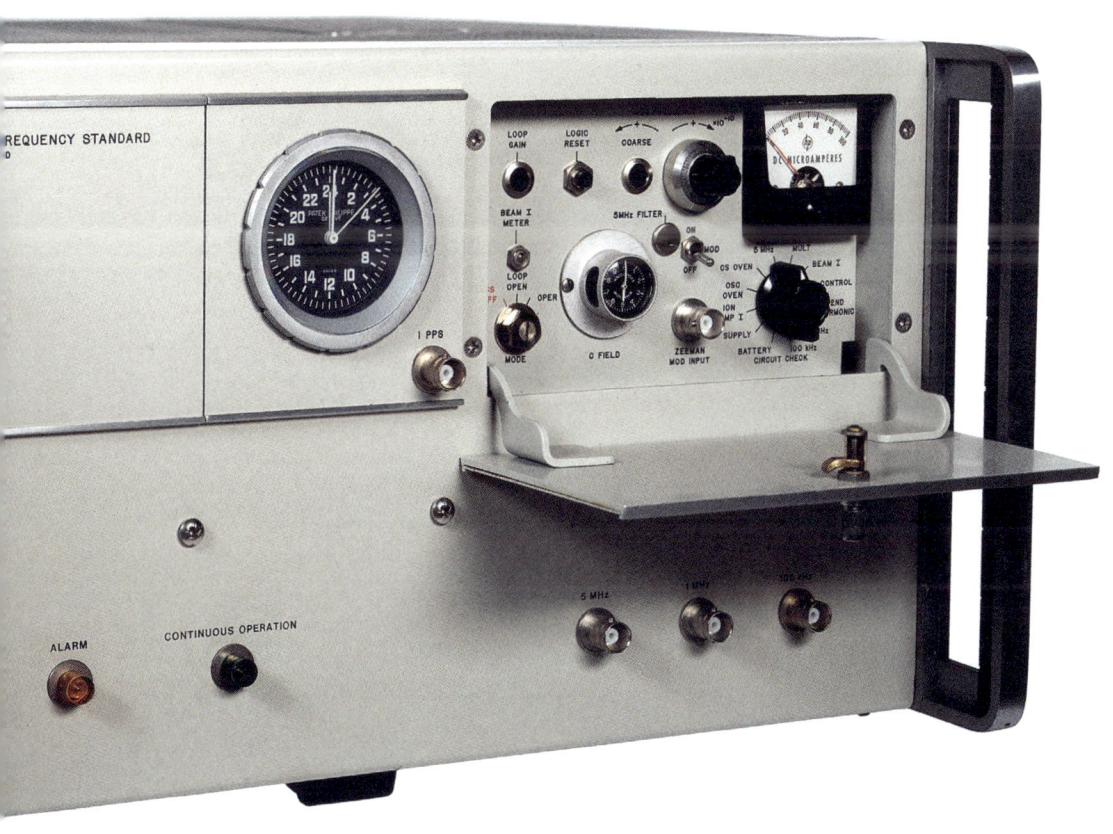

using new technologies to create highly reliable and portable commercial atomic clocks. The Observatory purchased its first HP5060A device in May 1966, followed by this model, HP5061A, a year later. At first, the Observatory combined its data with the NPL's to create a mean atomic clock, but by the mid-1970s it had assembled a set of six devices to set its own Greenwich Atomic Time. Scientists from the US Naval Observatory also transported their HP clocks to Herstmonceux to help check and calibrate the frequency standard.

Ultimately, no single observatory could be responsible for defining precise time in the atomic age. Multiple observatories and laboratories sent their data to the Bureau international de l'heure (BIH) in Paris for the new global time standard known as Coordinated Universal Time (UTC), which came into effect on 1 January 1972. One second was now officially defined as the time it takes for a caesium-133 atom to oscillate exactly 9,192,631,770 times. The Greenwich Time Service continued at Herstmonceux until February 1990 and Britain's time standards have since been defined by the latest atomic clocks at the NPL.

# Epilogue

In August 1948, Harold Spencer Jones departed Flamsteed House as the last Astronomer Royal to live and work at the Royal Observatory, Greenwich. It was the culmination of a ten-year search for a new observing location and the beginning of another decade of transition.

The question of leaving Greenwich had dogged the Astronomers Royal for over a century, with concerns raised about diminishing the Observatory's scientific authority, either by disrupting the continuity of long-term observations from a single location, or by losing access to the intellectual community of London's learned societies. Similarly, the Greenwich astronomers themselves were increasingly aware of how the site's identity and global status as the home of time and longitude were defined by its location.

But, by the 1930s, the site was no longer suitable for the new era of astrophysical observations with challenges emerging on multiple fronts: light pollution from electric street lighting, the tarnishing of telescope mirrors by nearby factory emissions and gritty atmospheric particles becoming trapped inside optical components. Spencer Jones used a combination of air pollution data and meteorological readings to state his case to the Admiralty but the decision to move was postponed by the outbreak of the Second World War and only approved on 17 February 1944. Ironically, the war itself provided more opportunities to relocate as several large country estates that had been requisitioned for wartime work became available after hostilities ceased. Spencer Jones reviewed over 70 possible sites before settling on the brick-built medieval castle and grounds at Herstmonceux, East Sussex. With clear open skies, good soils for drainage, minimal light pollution and nearby rail links to London, it was a highly pragmatic choice and the Astronomer Royal took up his new residence in the castle in October 1948.

## The Royal Greenwich Observatory (RGO), Herstmonceux

With its historic origins still firmly embedded within its title, the new Royal Greenwich Observatory, or 'RGO' as it became known, began operations in early 1949. The Solar Department was the first to relocate, with solar observations made with the same Dallmeyer photoheliograph at both Greenwich and Herstmonceux on a single day – 2 May 1949 – to ensure a continuation of the daily sunspot record (64, 72).

Chief Assistant Robert Atkinson was tasked with designing a set of six domes to house the equatorial telescopes due to be transferred from Greenwich, which also included an ingenious network of lifts and corridors that enabled technicians to transport safely the telescope mirrors from the domes into the maintenance workshop. Architect Brian O'Rorke was appointed to oversee the so-called 'Equatorial Group' but, with competing demands for his work on the 1951 Festival of Britain, construction of the domes dragged on until 1958. O'Rorke also had to work within strict planning requirements to minimise the visual impact of the new buildings. With this in mind, he specified telescope domes made of green copper in a bid to make the structures blend into the landscape, but the astronomers were exasperated to find that

Aerial view of Herstmonceux Castle,
Astral Aviation, 1930s
Graham Dolan

the metal domes trapped the warmth by day and lost heat at night, causing the air around the telescopes to shimmer. Similarly, a series of reflecting lily ponds and elevated walkways provided a decorative touch to the grounds but proved to be highly unsuitable for staff members walking between domes in the dark.

In contrast, the next phase of construction was overseen by the Admiralty's Civil Engineer-in-Chief, whose designs for the group of four transit sheds located to the north of the castle were much more utilitarian. Astronomers used these instruments to continue their observations for positional astronomy and timekeeping, mindful that they had to correct for the longitude difference between Greenwich and Herstmonceux to sustain the definition of the Prime Meridian and Greenwich Mean Time (GMT).

Finally, the area to the south-west of the castle became the site of a new purpose-built complex that was completed in 1957 and became known as the West Building. The sub-basement level contained the atomic clocks while the two upper levels provided offices, computing rooms, workshops and laboratories for the Nautical Almanac Office and Time Department.

At first, staff were provided with hostel-type accommodation in a series of surviving wartime huts near the castle but eventually the local council provided more permanent housing in local villages. The astronomers' enthusiasm for sports and social clubs (see *Sports and social clubs*)

View of the Equatorial Group domes postcard, unknown maker, about 1959
Graham Dolan

continued with the creation of a new association at Herstmonceux on 28 October 1948. At first, the activities were limited to billiards, snooker and table tennis but over time the list grew to include tennis, hockey, cricket and even country dancing in the castle's Long Gallery. Members worked hard to raise the funds for a new clubhouse that eventually opened on 1 October 1960.

### Changing roles
After 25 years' service, Spencer Jones retired in 1955, confident that the move to Herstmonceux was well underway. His successor, Richard van der Riet Woolley, adopted a very different approach to his predecessors with greater emphasis on astrophysical research in collaboration with academics, especially with the University of Sussex, and less focus on routine, long-term observations. The reallocation of the RGO's government funding from the Admiralty to the Science Research Council in 1965 reinforced its evolving status from government to academic institution and led to wider debates around the role of the Astronomer Royal. When Woolley retired in 1971, his management of the RGO was continued by newly appointed Director, Margaret Burbidge, while the title of Astronomer Royal became an honorary role bestowed on a leading British astronomer – in any field of astronomy, not just positional astronomy – with no requirement for any association with the RGO.

### Honorary Astronomers Royal
- 1972–82: Martin Ryle (University of Cambridge; radio astronomy)
- 1982–90: Francis Graham-Smith (University of Manchester; radio astronomy)
- 1991–95: Arnold Wolfendale (Durham University; cosmic rays)
- 1995–2025: Martin Rees (University of Cambridge; cosmology)
- 2025–present: Michele Dougherty (Imperial College London; planetary magnetospheres)

### Moving to Cambridge
As a subsidiary of the Science Research Council, the RGO now had to compete alongside universities for funding from an ever-decreasing pot of government funds. At the same time, the nature and emphasis of astronomy itself was changing. Bigger and better telescopes, funded by multinational consortia, were being constructed in the much clearer and more reliable skies of mountaintop locations in the

Canary Islands, Chile and Hawaii. Cheaper air travel also made it ever more feasible to send astronomers abroad rather than invest in new telescopes based in Britain. With repeated demands from the Treasury to reduce the RGO's costs and to cut staffing by half, the Observatory at Herstmonceux finally closed in 1990 and relocated again to the Institute of Astronomy at the University of Cambridge. The BBC took on responsibility for providing accurate time for its six pips signal (89) and the Isaac Newton Telescope – inaugurated at Herstmonceux in 1967 and famous for its role in the first detection of a black hole in 1971 – was transferred from the UK to the Roque de los Muchachos Observatory on La Palma in the Canary Islands. The new Cambridge RGO was mainly an administrative centre for Britain's involvement with overseas observatories and the RGO eventually ceased all operations in 1998.

### Changing definitions and standards

While astronomers turned their attention to more exotic and mysterious objects in the Universe, the everyday use of astronomy for navigation and timekeeping in the 1960s was being superseded by new technologies and the changing definition of the second. The adoption and renaming of Greenwich Mean Time as Universal Time (UT) for scientific, and later practical, purposes had already come into effect at international conferences held in 1884 (69) and 1928. By the 1930s, astronomers were using the enhanced accuracy of the Shortt freependulum clocks (90) to confirm their long-held suspicion that the Earth's rotation was variable and could not be relied upon as a timekeeping standard. Faced with this observational evidence, astronomers set about redefining the second. Rather than defining it as 1/86,400 of the mean solar day, based on the Earth's rotation, they created a new Ephemeris Time (ET) that defined a second as 1/31,556,925.9747 of the length of the year 1900, as measured from one spring equinox to the next. This definition was ratified in 1956 by the International Committee on Weights and Measures (known by its French acronym CPGM) and initially worked well for navigators and surveyors relying on the stars, but proved difficult to use in other contexts.

By the 1960s, scientists were starting to rely more on caesium atomic clocks (100), which provided a highly stable and uniform time standard that did not suffer from the variable effects of crystal 'ageing' seen in quartz clocks (99). Just over a decade after defining Ephemeris Time, scientists redefined the second again in 1967 as the duration of 9,192,631,770 transitions of a caesium-133 atom, which continues to be recognised as the official unit of time within the International System of Units (SI).

Yet despite these pivotal changes, astronomy remains part of our time standard. On 1 January 1972, scientists agreed to use a new standard called UTC, known in English as Coordinated Universal Time, which is composed of two parts, one atomic and one astronomical. For International Atomic Time (TAI), scientists at the Bureau international des poids et mesures (BIPM) in Paris collect data every five days from

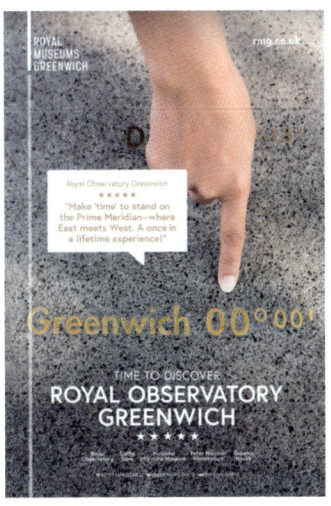

Tube poster featuring the buildings of the Royal Observatory, Warren Kenton, 1962
London Transport Museum

Campaign poster, 'Make 'time' to stand on the line', Royal Museums Greenwich, 2016 (L9851)

over 600 atomic clocks at around 90 locations worldwide, including data from the atomic clocks at the UK's National Physical Laboratory in Teddington. This atomic time is then compared to astronomical time (UT1) as supplied by the International Earth Rotation and Reference Systems Service (IERS) at Paris Observatory. Rather than relying on traditional meridian transits at observatories, astronomers can now measure the Earth's rotation using sophisticated techniques, such as combining radio signals from distant stars, firing lasers at low-level satellites to detect any variation in their orbits, or even using time signals from atomic clocks on various satellites for comparison.

Once the difference between atomic and astronomical time reaches 0.9 seconds, scientists add in an extra leap second to keep the time standards in sync. Since 1972 we have added 27 leap seconds to UTC, but this approach has become increasing problematic as we rely on computer networks running on atomic time for internet traffic, financial transactions and satellite navigation. Scientists are now exploring newer, more accurate atomic clocks that operate at higher frequency optical wavelengths – 100,000 times faster than the microwave frequencies in older atomic clocks – which may lead to revised definitions of the second within the next decade. In addition, BIPM members resolved in November 2022 to discontinue the use of leap seconds by 2035, meaning that the difference between atomic and astronomical time will not be adjusted until there is a significant discrepancy, possibly not for another 50 to 100 years.

**Where is the Prime Meridian?**
All these new timekeeping technologies and changing definitions of the second have affected the location of the Prime Meridian. In the 1970s, the US military developed a suite of satellites known as the Global Positioning System (GPS). Once a receiver on the ground had sight of at least four satellites above the horizon, it could use the time taken to receive the signals to calculate the distance between its location and the satellites, given that radio signals travel at the speed of light. By the 1980s, scientists realised that this new technology had implications for defining 0° longitude on Earth. The original Prime Meridian was defined by the Airy Transit Circle (47) and, like any measurement taken with an instrument on the Earth's surface, the resultant meridian was influenced by the planet's irregular shape and uneven crust, which cause local variations in gravity. In contrast, satellites orbiting the Earth's centre of mass are not susceptible to these localised effects and so we can use them to define our coordinate system over a much bigger scale, and use multiple satellites for greater accuracy.

In theory, scientists could have chosen any meridian plane passing through the Earth's centre as their zero reference but they decided to keep the new version as close as possible to Greenwich for continuity in measuring astronomical time for UTC. The satellite meridian was eventually defined as the plane passing 336ft (102.5m) east of the historic Airy meridian and was adopted as the International Reference Meridian (IRM) in 1984. It is the basis of our satellite navigation systems and time standards today and is maintained by a global network of ground stations – including one at Herstmonceux – that use laser-ranging to accurately monitor satellite positions.

**The historic site at Greenwich**
The Royal Observatory site was gradually absorbed within the responsibility of the National Maritime Museum in the 1950s and was opened to the public in stages, starting with the Octagon Room (2) in 1950, followed by the restoration of the Meridian Observatory in 1967. The Great Equatorial Telescope (75) was returned from Herstmonceux and installed in a new fibreglass version of the historic onion dome ready for the Observatory's tricentenary in 1975. Another landmark feature appeared in May 1993 with the installation of a green laser on the historic Prime Meridian. Titled *0 Degrees*, it is the work of artists Peter Fink and Anne Bean and can still be seen extending for over 30 miles (48km) today, although the construction of new apartment blocks in east London has necessitated a few tweaks to avoid projecting into people's homes!

As the home of time, the Observatory was a natural focal point for millennium celebrations on 31 December 1999, a moment in history denoted by an additional drop of the time ball at midnight. In 2007, the site was refurbished for the 'Time and Space' project, with new galleries and the construction of the Peter Harrison Planetarium, currently London's only planetarium.

Today the site attracts over 700,000 visitors each year and is undergoing a major new phase of

↑
A view of the laser projected across the Meridian Courtyard, 2024

(overleaf) *Prospectus intra Cameram Stellatam* [View inside the Star Room], Francis Place, 1712 (after Robert Thacker, 1676) (ZBA1808)

development known as 'First Light', to improve the displays, facilities and access for years to come.

**Legacy**
Our journey across three centuries and 100 objects has brought us to the end of the Royal Observatory's story but luckily the site's scientific legacy still continues through the work of specialist institutions such as the National Physical Laboratory, the Met Office, the British Geological Survey and the Hydrographic Office. Similarly, the Observatory's legacy lingers in our maps, technology and language as we check our position and change our watches across time zones. At first glance, the Observatory now seems redundant as we define time by atoms emitting light in laboratories and define location by constellations of satellites silently passing overhead. But we're still relying on the work of countless generations of Greenwich astronomers, technicians and human computers who worked tirelessly to turn a bewildering mass of observations into a global time and coordinate system. They were truly dedicated to the Observatory's founding principle of 'perfecting the art of navigation' and these intriguing objects will remind us of their stories for generations.

# Glossary

**achromatic** – a type of lens that does not disperse light into its constituent colours and minimises colour distortion.

**accuracy** – a measure of comparison between a clock and a known reference time, such as an observatory clock or atomic clock.

**altazimuth** – a type of telescope mount in which the instrument can be moved up and down (altitude) and rotated to any point along the horizon (azimuth).

**altitude** – the height (angle) of a celestial body above the horizon.

**aurorae** – glowing, swirling, coloured lights seen periodically in the night skies above the polar regions, known as the Northern Lights or Southern Lights within each hemisphere. They are caused by charged particles from the Sun interacting with the Earth's magnetic field.

**azimuth** – a point on the horizon that is described in degrees, starting at north as 0° or 360°. Also used to describe the horizontal movement of a telescope.

**celestial sphere** – a way of visualising the sky with the Earth at the centre of a hollow sphere covered with stars. Astronomers use imaginary circles and lines projected onto the sphere to measure star positions and time.

**circumpolar** – stars that never rise or set, as seen from a certain latitude. In the northern hemisphere they appear to circle around Polaris, the north star, every sidereal day.

**coelostat** – a mirror mounted on a motor that tracks the Sun's apparent path across the sky and constantly reflects sunlight into a nearby fixed telescope.

**constellation** – a group of stars in a defined region of sky. There are 88 formally recognised by the IAU constellations that are generally known by their historic names such as Orion (the Hunter) or Taurus (the Bull).

**corona** – the outermost layer of the Sun's atmosphere, generally seen during a total solar eclipse.

**declination (Dec)** – the celestial equivalent of latitude. Each star can be measured relative to the celestial equator, expressed as 0° to 90° north (+) or south (-).

**Doppler effect** – a shift in the frequency of waves depending on the motion of the source relative to an observer. A receding distant galaxy will appear to have its light shifted towards the red part of the spectrum (low frequency, long wavelength), while one approaching the Earth will appear to have light shifted towards the blue part of the spectrum (high frequency, short wavelength). Named after the Austrian scientist Christian Doppler (1803–53).

**ecliptic** – the Earth's orbital plane around the Sun, visible on the celestial sphere as the apparent path of the Sun, Moon and planets.

**equatorial** – a telescope mount in which the instrument moves in an arc parallel to the Earth's equator, making it easier to track objects as they appear to move across the sky.

**escapement** – the component of a mechanical clock that controls the speed at which the energy is released from the power source into the gear wheels, making the distinctive 'tick tock' sound and analogous to a valve, clutch, or a turnstile.

**going (of a clock)** – the time interval between successive windings of a clock or watch, usually expressed in days, e.g. a two- or eight-day marine chronometer.

**Greenwich Mean Time (GMT)** – a time standard originally based on observations of the Sun as seen at Greenwich. Natural variations in solar time caused by the Earth's elliptical orbit and tilted axis were smoothed out to create a regular, average (mean) length of time that could be used with clocks. GMT now refers to the name of the time zone used by certain countries in western Europe and Africa.

**horizontal magnetic force** – the angle between geographic north and magnetic north, which varies between locations and across time. Also known as magnetic declination or variation.

**latitude** – the measure of a location's north or south position, from 0° at the equator to 90° at the poles.

**local time** – the time at your location, as measured by observing the Sun crossing the local meridian at noon.

**longitude** – a location's east-west coordinate, measured from the Prime Meridian (0°) as +180° eastwards and -180° westwards.

**lunar distance method** – a navigational technique for finding longitude at sea whereby mariners use a sextant to measure the angle (lunar distance) between the outer edge (limb) of the Moon and a bright star, noting the time. After several calculations, mariners can compare this local observation to the time at which someone at a known location, such as Greenwich, would see the same angle in the sky. Since the Earth rotates 360° each day, a difference of one hour between observation means that the observing locations and reference location are separated by 15° longitude.

**magnetic intensity** – the strength of a magnetic field. Magnetic instruments respond most strongly where the field lines are the closest together.

**marine chronometer** – a portable, accurate timekeeper designed for use at sea to show the time at a reference location, such as Greenwich.

**meridian** – an imaginary line connecting the North and South Poles. The local meridian passes through the zenith at your location.

**micrometer** – an instrument used for making accurate measurements.

For astronomy, this usually involves having a series of vertical lines (wires) in the telescope eyepiece, similar to cross-hairs, to determine the exact moment at which a star crosses the meridian. Micrometers are also added to instruments for reading photographic plates to measure accurately sunspots and star positions.

**north star (Polaris)** – the brightest star in the constellation of Ursa Minor and the closest star to the celestial north pole. Its height (altitude) above the horizon is equivalent to your latitude.

**orbit** – the time it takes for a celestial body to make one complete revolution around its star or planet. The Earth takes 365.25 days to complete its orbit of the Sun.

**parallax** – the apparent shift in position of the Sun (solar parallax) or nearby stars (stellar parallax), as seen from different points on the Earth's orbit.

**photometry** – the measurement of the brightness or intensity of celestial objects.

**precision** – a measure of consistency and stability, meaning that a highly precise clock will have a similar rate all the time.

**Prime Meridian** – a meridian that is chosen to be 0° longitude for common reference. The Greenwich meridian was adopted as Prime Meridian of the World in 1884.

**prominences** – large looping structures that extend millions of miles/kilometres from the Sun's outer atmosphere. They are caused by the Sun's magnetic field and can last for days or months.

**proper motion** – the long-term shift in a star's position relative to the centre of mass of the Solar System.

**radial velocity** – how a star appears to move away or towards the Earth, as seen in the observer's line of sight.

**rate** – an indication of how much a clock will speed up or slow down, usually stated as a gain (+) or loss (-) in seconds per day. The smaller the rate, the more accurate the clock.

**reflecting telescope** – a type of telescope where light is collected and reflected by mirrors made from polished metal or glass coated with aluminium. Generally used at Greenwich for planetary observations and photographing faint objects such as gas clouds (nebulae) and galaxies.

**refracting telescope** – a type of telescope where light is collected and focused (refracted) by a series of lenses within an optical tube. Generally used at Greenwich for making transit observations, sketching the planets and taking systematic photographs of star fields.

**regulator** – a highly accurate pendulum clock used by astronomers at observatories with features to compensate for temperature variation and separate dials for hours, minutes and seconds for reading time more accurately.

**remontoire** – a secondary spring within a clock or watch that is periodically rewound automatically from the mainspring. This additional energy source helps provide a constant driving force to the escapement, making the timekeeper more accurate. The name stems from the French verb *remonter* (to wind).

**Right Ascension (RA)** – the celestial equivalent of longitude, measured relative to the vernal equinox. It is expressed in units of time from 0 to 24 hours with smaller divisions of minutes and seconds, increasing in value eastwards to match the Earth's rotation.

**sector** – an instrument used for measuring the angular separation between planets and stars.

**seeing** – an observer's visual assessment of turbulence in the Earth's atmosphere that might affect celestial observations.

**solar day** – the period of the Earth's rotation measured by successive crossings of the Sun on the local meridian at noon. A mean solar day is 24 hours long.

**sidereal day** – the period of the Earth's rotation measured by successive crossings of the vernal equinox on the local meridian. A mean sidereal day is 23 hours, 56 minutes and 4.09 seconds long.

**spectroscopy** – a technique for determining the chemical nature of materials in which emitted light is split via a prism to create a distinctive barcode-like pattern of lines for identification. Astronomers initially used spectroscopes to observe stellar spectral lines and later developed spectrographs to record the images, known as spectrograms.

**solstice** – the two moments each year when the Sun appears to reach the most extreme points of the ecliptic. Around 21 June it is directly overhead at noon as seen from the Tropic of Cancer, while around 21 December it is directly overhead at noon as seen from the Tropic of Capricorn, giving each hemisphere their longest and shortest days respectively.

**sunspots** – temporary dark patches on the surface of the Sun that signify areas of lower temperatures and higher magnetic intensities.

**transit** – the exact moment when a star crosses the local meridian.

**transit of Venus** – a rare astronomical alignment seen as a pair of events eight years apart which occurs approximately once a century. As the planet Venus appears to cross the surface of the Sun, astronomers can use the timings and angles of their observations to measure the scale of the Solar System.

**UTC** – Coordinated Universal Time (known as UTC) is the standard time by which the world regulates clocks and time. It is a combination of time measured by the Earth's daily rotation and time measured by multiple atomic clocks.

**vernal equinox** – a point on the celestial sphere used for measuring sidereal time, historically known as the First Point of Aries. The line drawn though this point is the celestial equivalent of the prime meridian and marks the start/end point of Right Ascension. The Sun crosses this point between the celestial equator and the ecliptic around 21 March each year.

**vertical magnetic force** – the angle between the Earth's surface and magnetic field lines, with a range of 0° at the equator up to 90° at the north magnetic pole. Also known as inclination or dip.

**zenith** – the point immediately above an observer, with an altitude of 90° above the horizon. The angle between a known star and the zenith is called the zenith distance and is used by astronomers, navigators and surveyors.

# Improvements in accuracy over time

The Observatory has relied on increasingly more accurate timekeepers over the centuries. For the first 250 years, astronomers measured the accuracy of their clocks in comparison with the Earth's rotation. Today, physicists rely on complex statistics to assess the accuracy of highly stable atomic clocks.

| OBJECT | DATE | TIMEKEEPER | OSCILLATOR | MEASURE OF 1 SECOND | TIME TO LOSE OR GAIN 1 SECOND (APPROX.) |
|---|---|---|---|---|---|
| 3 | 1675 | Tompion's year-going clock | 2-second pendulum | 0.5 swing | 1 second in 3 hours |
| 12 | 1725 | Regulator no.1 by Graham | 1-second pendulum | 1 swing | 1 second in 24 hours |
| 21 | 1759 | Harrison's fourth marine timekeeper, 'H4' | Spring balance | 5 beats | 1 second in 30 hours |
| 44 | 1840s | Regulator no.587 by Dent | 1-second pendulum | 1 swing | 1 second in 5 days |
| 90 | 1920s | Shortt free-pendulum clock system | 1-second pendulum in a partial vacuum | 1 swing | 1 second in 300 days |
| 99 | 1940s | Quartz clock and frequency standard | Quartz crystal excited by electricity | 100,000 vibrations or cycles | 1 second in 1,000 days |
| 100 | 1972 | HP caesium-beam atomic clock | Caesium-133 atoms excited by microwaves | 9,192,631,770 energy transitions | 1 second in 0.5 million years |
| Epilogue | 2025 | Optical atomic clock | Strontium atoms excited by lasers | 429,228,004,229,873 energy transitions | 1 second in 15 billion years |

# Additional readings, references and quotations

## GENERAL HISTORIES

- Andrewes, William J.H. (ed.), *The Quest for Longitude*, Harvard University Press, 1996
- Dolan, Graham and Devoy, Louise, *On the Line: The Story of the Greenwich Meridian*, National Maritime Museum, London, 2019
- Dunn, Richard and Higgitt, Rebekah, *Ships, Clocks and Stars: the quest for longitude*, HarperCollins, London, 2014
- Forbes, Eric, Meadows, Jack and Howse, Derek, *The Royal Observatory at Greenwich and Herstmonceux 1675–1975* (3 vols), Taylor and Francis, London, 1975
- Howse, Derek, *Greenwich Time and the Longitude*, Philip Wilson, London, 1997
- Maunder, E. Walter, *The Royal Observatory, Greenwich. A Glance at Its History and Work*, The Religious Tract Society, London, 1900

## BIOGRAPHICAL INFORMATION

- Clifton, Gloria and Turner, Gerard L'Estrange (ed.), *Directory of British Scientific Instrument Makers 1550–1851*, Zwemmer, London, 1995
- Hockey, Thomas, et al. *Biographical Encyclopaedia of Astronomers*, 2nd edition, Springer, New York, 2014
- O'Connor, J.J. and Robertson, E.F., MacTutor Biographies of Mathematicians, University of St Andrews, https://mathshistory.st-andrews.ac.uk
- Oxford Dictionary of National Biography (ODNB), https://oxforddnb.com/
- Winterburn, Emily, *The Astronomers Royal*, National Maritime Museum, London, 2003

## USEFUL WEBSITES

**British Geological Survey**
- https://geomag.bgs.ac.uk/operations/greenwich
- https://www.bgs.ac.uk/information-hub/scanned-records/magnetograms

**Grace's Guide**
- https://gracesguide.co.uk/Main_Page

**NASA Astrophysics Data System (ADS)**
- https://ui.adsabs.harvard.edu

**RGO Archives, Cambridge University Library**
- https://archivesearch.lib.cam.ac.uk/repositories/2/resources/88

**Royal Observatory Greenwich**
- http://www.royalobservatorygreenwich.org/articles.php

## REFERENCES, READINGS BY ENTRY AND QUOTATIONS ✢

### Introduction
- Airy, George Biddell, *Annual Report to the Board of Visitors*, 4 June 1853, p.8

### The quest for longitude
- de Grijs, Richard, 'A (not so) brief history of lunar distances: lunar longitude determination at sea before the chronometer', *Journal of Astronomical History and Heritage*, 23, 3, 2020, pp.495–522
- Williams, J., *From sails to satellites: the origin and development of navigational science*, Oxford University Press, Oxford, 1994, chapters 6 and 7
- ✢ National Archives, Kew, Royal Warrant, 4 March 1675, State Papers Domestic SP 44/334, p.10
- ✢ Frisius, Gemma, *De Principiis Astronomiae Cosmographicae*, Louvain, 1530, Chapter 19, 'Concerning a new method of finding longitude'; see English translation by Philip Kay in Howse, Derek, *Greenwich Time and the Longitude*, Philip Wilson, London, 1997, p.24

### John Flamsteed, first Astronomer Royal
- Baily, Francis, *An Account of the Reverend John Flamsteed*, Lords Commissioners of the Admiralty, London, 1835
- Flamsteed, John, *The correspondence of John Flamsteed, the first astronomer royal, 1666–1719*, Forbes, E.G., Murdin, L. and Wilmoth, F. (eds), 3 vols, Institute of Physics, Bristol, 1995–2002
- ✢ National Archives, Kew, Royal Warrant, 4 March 1675, State Papers Domestic SP 44/334, p.10
- ✢ Brück, Mary, *Women in Early British and Irish Astronomy. Stars and Satellites*, Springer Netherlands, 2009, p.3
- ✢ Flamsteed, John, *The correspondence of John Flamsteed, the first astronomer royal, 1666–1719*, Forbes, E.G., Murdin, L. and Wilmoth, F. (eds), 3 vols, Institute of Physics, Bristol, 1995–2002, letter 727, Samuel Pepys to John Flamsteed, 5 July 1697, p.651

### 1. Portrait of Charles II
- Howse, Derek, *Greenwich Time and the Longitude*, Philip Wilson, London, 1997, pp.33–51
- Forbes, Eric, *Greenwich Observatory: The Royal Observatory at Greenwich and Herstmonceux 1675–1975, Volume 1: Origins and Early History (1675–1835)*, Taylor and Francis, London, 1975, pp.18–24
- ✢ National Archives, Kew, Royal Warrant, 4 March 1675, State Papers Domestic SP 44/334, p.10

### 2. Print of *Prospectus intra Cameram Stellatam*
- Van Helden, Albert, 'The Telescope in the Seventeenth Century', *Isis*, 65, 1, 1974, pp.38–58
- Howse, Derek, *Francis Place and the early history of the Greenwich Observatory*, Science History Publications, New York, 1975

### 3. Year-going clock movement
- Cescinsky, Herbert, 'Flamsteed's Clocks from Greenwich Observatory', *The Burlington Magazine*, 51, 297, 1927, pp.309–11
- Howse, Derek, *The Tompion Clocks at Greenwich and the Dead-beat Escapement*, The Antiquarian Horological Society, London, 1970

### 4. Flamsteed's 7ft equatorial sextant
- Chapman, Alan, *Dividing the circle: the development of critical angle measurement in Astronomy, 1500–1850* (2nd edition), John Wiley and Sons in association with Praxis Publishing, Chichester, 1995, chapter 3
- Howse, Derek, *Greenwich Observatory: The Royal Observatory at Greenwich and Herstmonceux 1675–1975, Volume 3: The Buildings and Instruments*, Taylor and Francis, London, 1975, pp.75–9

### 5. The Royal Observatory from Crooms Hill
- Howse, Derek, *Greenwich Observatory: The Royal Observatory at Greenwich and Herstmonceux 1675–1975, Volume 3: The Buildings and Instruments*, Taylor and Francis, London, 1975, pp.58–60, 109–10
- King, Henry C., *The History of the Telescope*, Dover Publications Inc., New York, 1955, chapter 4, p.6
- ✢ Forbes, Eric, Meadows, Jack and Howse, Derek, *The Royal Observatory at Greenwich and Herstmonceux 1675–1975*, Taylor and Francis, London, 1975, Volume III: Buildings and Instruments, p.110

### 6. 1714 Longitude Act
- Howse, Derek, *Greenwich Time and the Longitude*, Philip Wilson, London, 1997, pp.55–8
- UK Parliamentary Archives, 'An Act for providing a Public Reward for such Person or Persons as shall discover the Longitude at Sea', GB-061, Public Acts, 13 Anne, c.14, 1713
- ✢ National Archives, Kew, 'An Act for providing a Public Reward for such Person or Persons as shall discover the Longitude at Sea', GB-061, Public Acts, 13 Anne, c.14, 1713

### 7. Historia Coelestis Britannica (1725)
- Baily, Francis, *An Account of the Reverend John Flamsteed*, Lords Commissioners of the Admiralty, London, 1835, Part II: Containing the British Catalogue
- Howse, Derek, *Greenwich Time and the Longitude*, Philip Wilson, London, 1997, pp.49–51
- Kanas, Nick, *Star Maps: History, Artistry, and Cartography*, Springer Praxis, Chichester, 2007, pp.172–5
- ✢ Brewster, David, *Memoirs of the Life, Writings, and Discoveries of Sir Isaac Newton, Volume 2*, T. Constable and Company, Edinburgh, 1855, p.157
- ✢ Flamsteed, John (1684), *The correspondence of John Flamsteed, the first astronomer royal, 1666–1719*, Forbes, E.G., Murdin, L. and Wilmoth, F. (eds), 3 vols, Institute of Physics, Bristol, 1995–2002. Letter 1424, John Flamsteed to Abraham Sharp, 2 June 1716, p.795
- ✢ Baily, Francis, *An Account of the Revd. John Flamsteed, the First Astronomer-royal*, London, 1835, 'Seventh Division, From 1704 to 1716', p.101
- ✢ National Archives, Kew, Royal Warrant, 4 March 1675, State Papers Domestic SP 44/334, p.10

### 8. Atlas Coelestis (1729)
- Blitzstein, W., 'The seven identified observations of Uranus made by John Flamsteed using his mural arc', *The Observatory*, 118, 1998, pp.219–22
- Kanas, Nick, *Star Maps: History, Artistry, and Cartography*, Springer Praxis, Chichester, 2007, pp.172–5
- ✢ Baily, Francis, *An Account of the Revd. John Flamsteed, the First Astronomer-royal*, Lords Commissioners of the Admiralty, London, 1835, p.363

### Edmond Halley, second Astronomer Royal
- Cook, Alan, *Edmond Halley: Charting the Heavens and the Seas*, Clarendon Press, Oxford, 1998
- Love, David, *Edmond Halley: The Many Discoveries of the Most Curious Astronomer Royal*, Prometheus, Amherst NY, 2023

### 9. Chart of the southern celestial hemisphere
- Cook, Alan, *Edmond Halley: Charting the Heavens and the Seas*, Clarendon Press, Oxford, 1998, pp.60–79
- Ridpath, Ian, 'Star Tales: Edmond Halley's southern star catalogue: First accurate survey of the southern sky', http://www.ianridpath.com/startales/halley.html (accessed 21 December 2023)

### 10. Halley's transit telescope
- Fearon, S. and Eyes, J., *A Description of the Sea Coast of England and Wales*, A. Sadler, Liverpool, 1738
- Howse, Derek, *Greenwich Observatory: The Royal Observatory at Greenwich and Herstmonceux 1675–1975, Volume 3: The Buildings and Instruments*, Taylor and Francis, London, 1975, pp.31–4

### 11. Halley's 8ft mural quadrant
- Cook, Alan, *Edmond Halley: Charting the Heavens and the Seas*, Clarendon Press, Oxford, 1998, pp.394–5
- Howse, Derek, *Greenwich Observatory: The Royal Observatory at Greenwich and Herstmonceux 1675–1975, Volume 3: The Buildings and Instruments*, Taylor and Francis, London, 1975, pp.31–4

### How does a mechanical clock work?
- Graham, George, 'IV. A contrivance to avoid the irregularities in a clocks motion, occasion'd by the action of heat and cold upon the rod of the pendulum', *Philosophical Transactions of the Royal Society of London*, 34, 392, 1726, pp.40–4

### 12. Regulator no.1 by Graham
- Harrison, John, *A Description Concerning Such Mechanism as will Afford a Nice, or True Mensuration of Time*, 1775, National Maritime Museum Archives, ref. REC/71, f.21
- Hellman, C. Doris, 'George Graham: maker of horological and astronomical instruments', *Popular Astronomy*, 39, 1931, pp.186–98
- McEvoy, R., 'George Graham and the Orrery', *Nuncius*, 35, 2, 2020, pp.235–50

### 13. Marine timekeeper by Harrison, 'H1'
- Betts, Jonathan, *Marine Chronometers at Greenwich*, Oxford University Press, Oxford, 2017, pp.18–20, 134–47
- Betts, Jonathan, *John Harrison and the Quest for Longitude* (2nd edition), National Maritime Museum, London, 2023, pp.53–9
- ✢ Sir Charles Wagner to Captain Proctor, 14 May 1736, quoted in John Harrison, *An Account of the Proceedings, in Order to the Discovery of the Longitude*, London, 1763, p.17

### 14. Marine timekeeper by Harrison, 'H2'
- Betts, Jonathan, *Marine Chronometers at Greenwich*, Oxford University Press, Oxford, 2017, pp.20–1, 147–60
- Betts, Jonathan, *John Harrison and the Quest for Longitude* (2nd edition), National Maritime Museum, London, 2023, pp.60–4
- RGO Archives, 'Confirmed minutes of the Board of Longitude, 1737–1779', RGO 14/5, https://cudl.lib.cam.ac.uk/view/MS-RGO-00014-00005/1 (accessed 21 December 2023)
- ✢ Board of Longitude meeting, 16 January 1741, 'Confirmed minutes of the Board of Longitude, 1737–1779', RGO Archives, RGO 14/5, https://cudl.lib.cam.ac.uk/view/MS-RGO-00014-00005/1 (accessed 22 August 2024)

### James Bradley, third Astronomer Royal
- Fisher, John, *The Life and Work of James Bradley: The New Foundations of 18th Century Astronomy*, Oxford University Press, 2024
- Hirshfeld, A.W., 'Bradley, James', in Hockey, Thomas, et al. *The Biographical Encyclopedia of Astronomers*, Springer, New York, 2007
- Murdin, P., 'Molyneux, Samuel', in Hockey, Thomas, et al. *The Biographical Encyclopedia of Astronomers*, Springer, New York, 2007

### 15. Bradley's 12.5ft zenith sector
- Howse, Derek, *Greenwich Observatory: The Royal Observatory at Greenwich and Herstmonceux 1675–1975, Volume 3: The Buildings and Instruments*, Taylor and Francis, London, 1975, pp.60–5
- Walsh, K.A.P., 'James Bradley and reflections on a special year', *Astronomy & Geophysics*, 53, 6, 2012, pp.6.31–6.33

### 16. Bradley's 8ft mural quadrant
- Bird, John, *The Method of Constructing Mural Quadrants Exemplified by a Description of the Brass Mural Quadrant in the Royal Observatory at Greenwich*, London, W. Richardson and S. Clark, 1768
- Howse, Derek, *Greenwich Observatory: The Royal Observatory at Greenwich and Herstmonceux 1675–1975, Volume 3: The Buildings and Instruments*, Taylor and Francis, London, 1975, pp.24–6
- Rigaud, S.P., *Miscellaneous Works and Correspondence of the Rev. James Bradley, D.D., F.R.S.*, Oxford University Press, 1832, pp.lxxvi–viii, pp.84–5

### 17. Bradley's transit telescope
- Howse, Derek, *Greenwich Observatory: The Royal Observatory at Greenwich and Herstmonceux 1675–1975, Volume 3: The Buildings and Instruments*, Taylor and Francis, London, 1975, pp.34–8

### 18. Marine timekeeper by Harrison, 'H3'
- Betts, Jonathan, *Marine Chronometers at Greenwich*, Oxford University Press, Oxford, 2017, pp.21–3, 160–74
- Betts, Jonathan, *John Harrison and the Quest for Longitude* (2nd edition), National Maritime Museum, London, 2023, pp.67–72
- RGO Archives, 'Confirmed minutes of the Board of Longitude, 1737–1779', RGO 14/5, https://cudl.lib.cam.ac.uk/view/MS-RGO-00014-00005/1 (accessed 21 December 2023)
- Board of Longitude meeting, 4 June 1746, 'Confirmed minutes of the Board of Longitude, 1737–1779', RGO Archives, RGO 14/5, https://cudl.lib.cam.ac.uk/view/MS-RGO-00014-00005/1 (accessed 22 August 2024)

### Nathaniel Bliss, fourth Astronomer Royal
- Bliss, Nathaniel, 'XXXII. Observations on the transit of Venus over the sun, on the 6th of June 1761', *Philosophical Transactions of the Royal Society of London*, 52, 1761, pp.173–77
- Fauvel, John, Flood, Raymond and Wilson, Robin, *Oxford Figures: Eight Centuries of the Mathematical Sciences*, Oxford University Press, 2013, pp.177–8
- ✢ National Archives, Kew, Royal Warrant, 4 March 1675, State Papers Domestic, SP 44/334, p.10

### Nevil Maskelyne, fifth Astronomer Royal
- Higgitt, Rebekah (ed.), *Maskelyne: Astronomer Royal*, Robert Hale Ltd, London, 2014
- Howse, Derek, *Nevil Maskelyne: The Seaman's Astronomer*, Cambridge University Press, 1989

### 19. *Nautical Almanac* (1767)
- Admiralty, 'The Nautical Almanac (NP314)', https://www.admiralty.co.uk/publications/astronomical-publications/the-nautical-almanac (accessed 14 March 2025)
- Croarken, Mary, 'Providing longitude for all', *Journal for Maritime Research*, 4, 1, 2002, pp.106–26
- Dunn, Richard and Higgitt, Rebekah, *Ships, Clocks and Stars: The Quest for Longitude*, HarperCollins, London, 2014, pp.109–14
- Forbes, E., 'The Foundation and Early Development of the Nautical Almanac' *Journal of Navigation*, 18, 4, 1965, pp.391–401
- RGO Archives, 'An Act [1765, 5 Geo 3. c.20] for explaining and rendering more effectual Two Acts … with regard to the making of Experiments of Proposals made for the discovery of the Longitude; and to enlarge the Number of Commissioners for putting in Execution the Said Act', RGO14/1/29

### 20. Maskelyne's observing suit
- Croarken, Mary, 'Astronomical labourers: Maskelyne's assistants at the Royal Observatory, Greenwich, 1765–1811,' *Notes and Records of the Royal Society*, 57, 3, 2003, pp.285–98
- Evans, John, *Juvenile tourist*, James Cundee, London, 1810, pp.333–5
- Miller, Amy, 'The Maskelynes at home', in *Maskelyne: Astronomer Royal*, Rebekah Higgitt (ed.), Robert Hale Ltd, London, 2014
- ✢ Evans, John, *Juvenile tourist*, James Cundee, London, 1810, p.334

### 21. Marine timekeeper by Harrison, 'H4'
- Betts, Jonathan, *Marine Chronometers at Greenwich*, Oxford University Press, Oxford, 2017, pp.31–4, 174–85
- Betts, Jonathan, *John Harrison and the Quest for Longitude* (2nd edition), National Maritime Museum, London, 2023, pp.86–90
- Maskelyne, Nevil, 'An account of the going of Mr. John Harrison's watch, at the Royal Observatory, from May 6th, 1766, to March 4th, 1767, The Commissioners of Longitude, London, 1767
- RGO Archives, 'Confirmed minutes of the Board of Longitude, 1737–1779', RGO 14/5, https://cudl.lib.cam.ac.uk/view/MS-RGO-00014-00005/1 (accessed 21 December 2023)
- RGO Archives, 'Trials of Harrison's chronometers, 1766–1771', RGO 4/311, https://cudl.lib.cam.ac.uk/view/MS-RGO-00004-00311/1 (accessed 22 December 2023)
- ✢ Maskelyne, Nevil, 'An account of the going of Mr. John Harrison's watch, at the Royal Observatory, from May 6th, 1766, to March 4th, 1767, The Commissioners of Longitude, London, 1767, p.24

### 22. Pair of floor globes
- RGO Archives, 'Cover letter for a celestial globe recently sent to Andrews, 1810-03-19', RGO 4/149, f.75
- RGO Archives, 'Letters from Maskelyne to Henry Andrews, 1768–1811', RGO 4/149, f.6
- RGO Archives, 'Printers' and publishers' accounts, 1782–1829', RGO 14/16, f.438r
- Dunn, Richard and Higgitt, Rebekah, *Ships, Clocks and Stars: The Quest for Longitude*, HarperCollins, London, 2014, pp.112–14

### 23. Longitude calculation sheet
- Bishop, Robert, *Instructions and observations relative to the navigation of the Windward and Gulph passages*, Robert Bishop, London, 1766, Introduction and Appendix
- Croarken, Mary, 'Providing longitude for all', *Journal for Maritime Research*, 4, 1, 2002, pp.106–26
- Miller, David, 'Longitude Networks on Land and Sea', in *Navigational Enterprises in Europe and Its Empires, 1730–1850*, Richard Dunn and Rebekah Higgitt (eds), Palgrave Macmillan, London, 2016, pp.232–4
- ✢ Bishop, Robert, *Instructions and observations relative to the navigation of the Windward and Gulph passages*, Robert Bishop, London, 1766, title page

### Transits of Venus expeditions
- Airy, George Biddell, *Account of observations of the transit of Venus, 1874, December 8, made under the authority of the British government, and of the reduction of the observations*, HMSO, London, 1881
- Exploratorium, San Francisco, 'The rarest eclipse: Transit of Venus', https://annex.exploratorium.edu/venus/index.html (accessed 14 March 2025)
- Hughes, David W., 'Six stages in the history of the astronomical unit', *Journal of Astronomical History and Heritage*, 4, 1, 2001, pp.15–28
- International Astronomical Union, 'Measuring the Universe: The IAU and astronomical units', https://www.iau.org/public/themes/measuring/ (accessed 27 December 2023)
- Orchiston, Wayne, 'Cook, Green, Maskelyne and the Transit of Venus 1769: the legacy of the Tahitian

observations', *Journal of Astronomical History and Heritage*, 20, 1, 2017, pp.55–68
- Wulf, Andrea, *Chasing Venus: The race to measure the heavens*, Windmill Books, London, 2013
✣ *Abridged Transactions of the Royal Society*, Volume VI, 1809, pp.243–9. Available at https://eclipse.gsfc.nasa.gov/transit/HalleyParallax.html (accessed 22 August 2024)
✣ Airy, George Biddell, *Account of observations of the transit of Venus, 1874, December 8*, HMSO, London, 1881, p.viii

### 24. Gregorian telescope
- Green, Charles and Cook, James, 'XLIII. Observations made, by appointment of the Royal Society, at King George's Island in the South Sea', *Philosophical Transactions of the Royal Society of London*, 61, 1771, pp.397–421
- Gee, Brian, *Francis Watkins and the Dollond Telescope Patent Controversy*, McConnell, Anita and Morrison-Low, Alison (eds), Ashgate Publishing, 2014
- King, Henry C., *The History of the Telescope*, Dover Publications Inc., New York, 1955
- Orchiston, Wayne, 'Cook, Green, Maskelyne and the Transit of Venus 1769: the legacy of the Tahitian observations', *Journal of Astronomical History and Heritage*, 20, 1, 2017, pp.35–68
- Royal Society Archives, 'Minutes of a meeting of the Council of the Royal Society, 12 November 1767', CMO/5/92
✣ Green, Charles and Cook, James, 'XLIII. Observations made, by appointment of the Royal Society, at King George's Island in the South Sea', *Philosophical Transactions of the Royal Society of London*, 61, 1771, p.411

### 25. Maskelyne's Copley Medal
- Danson, Edwin, *Weighing the world: the quest to measure the Earth*, Oxford University Press, New York, 2006, chapter 13
- Higgitt, Rebekah, '"In the Society's Strong Box": A Visual and Material History of the Royal Society's Copley Medal, c. 1736–1760', *Nuncius*, 34, 2, 2019, pp.284–316
- Maskelyne, Nevil, 'XLVIII. A proposal for measuring the attraction of some hill in this kingdom by astronomical observations', *Philosophical Transactions of the Royal Society of London*, 65, 1775, pp.495–500
- Maskelyne, Nevil, 'XLIX. An account of observations made on the mountain Schehallien for finding its attraction', *Philosophical Transactions of the Royal Society of London*, 65, 1775, pp.500–42
- Royal Society, 'Copley Medal', https://royalsociety.org/grants-schemes-awards/awards/copley-medal/ (accessed 28 December 2023)
- Smallwood, John R., 'Maskelyne's 1774 Schiehallion experiment revisited', *Scottish Journal of Geology* 43, 2007, pp.15–31
✣ Royal Society, 'Copley Medal' https://royalsociety.org/medals-and-prizes/copley-medal/ (accessed 22 August 2024)

### 26. Chronometer no.36 by Arnold
- Arnold, John, *An Account kept during Thirteen Months in the Royal Observatory at Greenwich of The going of a Pocket Chronometer, Made on a new Construction…*, John Arnold, London, 1780
- Betts, Jonathan, 'Arnold & Earnshaw: The Practicable Solution', in *The Quest for Longitude*, William Andrewes (ed.), Harvard University Press, 1996, pp.311–28
- Betts, Jonathan, *Marine Chronometers at Greenwich*, Oxford University Press, 2017, p.48
- RGO Archives, 'Confirmed Minutes of the Board of Longitude 6th March 1779', RGO 14/5, f.343v, https://cudl.lib.cam.ac.uk/view/MS-RGO-00014-00005/346 (accessed 14 March 2025)
✣ Arnold, John, *An Account kept during Thirteen Months in the Royal Observatory at Greenwich of The going of a Pocket Chronometer, Made on a new Construction…*, John Arnold, London, 1780, p.6
✣ Davidson, Simon C., 'Marine chronometers: the rapid adoption of new technology by East India captains in the period 1770–1792 on over 580 voyages', *Antiquarian Horology*, 40, 1, 2019, p.84

### 27. Sextant by Ramsden
- Mörzer Bruyns, W.F.J. and Dunn, Richard, *Sextants at Greenwich*, Oxford University Press, 2009
- McConnell, Anita, *Jesse Ramsden (1735–1800): London's Leading Scientific Instrument Maker*, Routledge, 2007
- Stimson, A. '18. The influence of the Royal Observatory at Greenwich upon the design of 17th and 18th century angle-measuring instruments at sea', *Vistas in Astronomy*, 20, 1976, pp.123–30

### London's community of clock and scientific instrument makers, 1750–1800
- Clifton, Gloria and Turner, Gerard L'Estrange (ed.), *Directory of British Scientific Instrument Makers 1550–1851*, Zwemmer, London, 1995, pp.xi–xv
- Desborough, Jane and Clifton, Gloria, 'Science and the City: The role of women in the science city: London 1650–1800', *Science Museum Group Journal*, 15, Spring 2021, https://dx.doi.org/10.15180/211502/001 (accessed 14 March 2025)
- Ginn, William T., 'Philosophers and artisans: the relationship between men of science and instrument makers in London 1820–1860', PhD diss., University of Kent at Canterbury, 1991
- Kilburn-Toppin, Jasmine, 'Instrument makers, shops, and expertise in eighteenth-century London', in *Spaces of Enlightenment Science, Knowledge Infrastructure and Knowledge Economy*, Gordon McOuat and Larry Stewart (eds), Brill, 2022, pp.74–90
- Morrison-Low, Alison, *Making Scientific Instruments in the Industrial Revolution*, Routledge, 2007
- RGO Archives, 'Memorandum book for 1801–1806, 1801–1806' [Maskelyne's food accounts], RGO 218/2/10
- RGO Archives, 'Memorandum book 'Directions', 1804–1810' [Maskelyne's record of dinner parties], RGO 218/2/12

### 28. Sophia Maskelyne's dress
- Herschel, Caroline and Herschel, Mary (ed.), *Memoir and Correspondence of Caroline Herschel*, John Murray, London, 1876; letter from Sophia dated 30 August 1799
- Miller, Amy, 'The Maskelynes at home', in *Maskelyne: Astronomer Royal*, Rebekah Higgitt (ed.), Robert Hale Ltd, London, 2014
✣ Herschel, Caroline and Herschel, Mary (ed.), *Memoir and Correspondence of Caroline Herschel*, John Murray, London, 1876; letter from Sophia dated 30 August 1799, p.101

### 29. Portrait of Margaret Maskelyne
- Croarken, Mary, 'Astronomical labourers: Maskelyne's assistants at the Royal Observatory, Greenwich, 1765–1811', *Notes and Records of the Royal Society*, 57, 3, 2003, pp.285–98
- Miller, Amy, 'The Maskelynes at home', in *Maskelyne: Astronomer Royal*, Rebekah Higgitt (ed.), Robert Hale Ltd, London, 2014
- RGO Archives, 'Memorandum book, 'Memoranda July 5th 1782 to July 24th 1788', 1782–1806', RGO 218/2/5, f.19v
✣ RGO Archives, 'Memorandum book, 'Memoranda July 5th 1782 to July 24th 1788', 1782–1806', RGO 218/2/5, f.19v

### 30. Caroline Herschel's dress and bonnet
- Brück, Mary, *Women in Early British and Irish Astronomy: Stars and Satellites*, Springer Science and Business Media, 2009
- Herschel, Caroline and Herschel, Mary (ed.), *Memoir and Correspondence of Caroline Herschel*, John Murray, London, 1876
✣ Herschel, Caroline and Herschel, Mary (ed.), *Memoir and Correspondence of Caroline Herschel*, John Murray, London, 1876, pp.37, 41, 47, 89, 102

**31. *Astronomical Observations* (1798)**
- Forbes, Eric, 'Dr. Bradley's Astronomical Observations', *Quarterly Journal of the Royal Astronomical Society*, 6, 1965, pp.321-28
- Maskelyne, Nevil, 'Concerning the Latitude and Longitude of the Royal Observatory at Greenwich; With Remarks on a Memorial of the Late M. Cassini de Thury', *Philosophical Transactions of the Royal Society of London*, 77, 1787, pp.151-87

**32. Model of a chronometer escapement**
- Banks, Joseph, *Protest against a vote of the Board of longitude, granting to Mr. Earnshaw a reward for the merit of his time-keepers*, W. Bulmer, London, 1804
- Betts, Jonathan, *Marine Chronometers at Greenwich*, Oxford University Press, 2017, pp.50-7
- Dunn, Richard and Higgitt, Rebekah, *Ships, Clocks and Stars: The Quest for Longitude*, HarperCollins, London, 2014, pp.166-72
- Maskelyne, Nevil, *Arguments for giving a reward to Mr Earnshaw for his improvements on timekeepers*, Luke Hansard, London, 1804
- McEvoy, R. 'Maskelyne's time', in *Maskelyne: Astronomer Royal*, Rebekah Higgitt (ed.), Robert Hale Ltd, London, 2014
- RGO Archives, 'Confirmed minutes of the Board of Longitude, 1802-1823', RGO Archives, RGO 14/7, https://cudl.lib.cam.ac.uk/view/MS-RGO-00014-00007/1 (accessed 3 January 2024)

**33. Print of the Easter Festival**
- Dickens, Charles, 'Scenes: Chapter 12: Greenwich Fair', *Sketches by "Boz"*, John Macrone, London, 1836; originally published as 'Sketches of London No. 9', *The Evening Chronicle*, 16 April 1835
- Longhurst, Ronald, 'Greenwich Fair', *Transactions of the Greenwich and Lewisham Antiquarian Society*, 7, 4, 1970, pp.198-210
- Walford, Edward, 'Greenwich: The park and the royal observatory', *Old and New London, Volume 6*, Cassell, Petter & Galpin, London, 1878, pp.206-23
- ✦ Dickens, Charles, 'Scenes: Chapter 12: Greenwich Fair', *Sketches by "Boz"*, John Macrone, London, 1836; originally published as 'Sketches of London No. 9', *The Evening Chronicle*, 16 April 1835

**34. Print of the view from the camera obscura**
- Brennan, Pip, *The Camera Obscura and Greenwich*, National Maritime Museum, London, 1994
- Flamsteed, John, *The correspondence of John Flamsteed, the first astronomer royal, 1666-1719*, Forbes, E.G., Murdin, L. and Wilmoth, F. (eds), 3 vols, Institute of Physics, Bristol, 1995-2002; letter 519, from Flamsteed to Bernard, 8 July 1684, pp.176-7
- Hammond, John H., *The Camera Obscura: A Chronicle*, Taylor & Francis, United Kingdom, 1981
- Howse, Derek, *Francis Place and the Early History of the Greenwich Observatory*, Science History Publications, New York, 1975, plate Xa, 'Domus Obscurata'
- Hobhouse, Hermione (ed.), 'Chapter XIII: The Isle of Dogs: Introduction', *Survey of London: Volumes 43 and 44, Poplar, Blackwall and Isle of Dogs*, London County Council, 1994, pp.375-87, https://www.british-history.ac.uk/survey-london/vols43-4/ (accessed 4 January 2024)

**John Pond, sixth Astronomer Royal**
- Dolan, Graham, 'People: John Pond, the sixth Astronomer Royal', http://www.royalobservatorygreenwich.org/articles.php?article=1301 (accessed 27 December 2024)
- ✦ Royal Society Archives, 'Greenwich Observatory Papers', MS/371, volume XXX, f.49

**35. Pond's mural circle**
- Howse, Derek, *Greenwich Observatory: The Royal Observatory at Greenwich and Herstmonceux 1675-1975, Volume 3: The Buildings and Instruments*, Taylor and Francis, London, 1975, pp.26-9
- Pond, John, 'Explanation of the Method of Observing with the Two Mural Circles, as Practised at Present at the Royal Observatory of Greenwich', *Astronomical Observations made at the Royal Observatory at Greenwich*, 11, 1826, N8-N9

**36. Astronomer's alarm clock**
- Dolan, Graham, 'People: Thomas Taylor', http://www.royalobservatorygreenwich.org/articles.php?article=1125 (accessed 27 December 2024)
- Dolan, Graham, 'The Airy Transit Circle' [explanation of clock stars], http://www.thegreenwichmeridian.org/tgm/articles.php?article=6 (accessed 27 December 2024)
- ✦ RGO Archives, 'Admiralty correspondence, 1835-1844', RGO6/72, f.223 and f.226

**37. Rain gauge**
- Airy, George Biddell, 'Rain gauges', *Magnetic and Meteorological Observations at the Royal Observatory, Greenwich, for the year 1840*, Palmer and Clayton, London, 1843, p.lxxxii
- Nash, W.C., 'Monthly rainfall at the Royal Observatory, Greenwich, 1815-1903', *Quarterly Journal of the Royal Meteorological Society*, 30, 1904, pp.291-306
- RGO Archives, 'Daily weather records, 1807-01-01-1821-10', RGO 5/241

**38. Chronometer no.427 by Brockbanks**
- Airy, George Biddell, 'Eclipses, Occultations, and Transits of Jupiter's Satellites…1864', *Greenwich Observations*, 1866, p.68
- Betts, Jonathan, *Marine Chronometers at Greenwich*, Oxford University Press, 2017, pp.211-13
- Stewart, A.D., 'John & Miles Brockbank, their life and work', *Antiquarian Horology*, 36, 4, 2015, pp.502-20
- ✦ RGO Archives, 'Admiralty chronometer six day book, 1821-1827', letter from John Pond to Francis Beaufort, 27 November 1829, RGO 5/232, f.51

**39. Time ball**
- Bartky, Ian R. and Dick, Steven J., 'The First Time-Balls', *Journal for the History of Astronomy*, 12, 3, 1981, pp.155-64
- Bateman, Doug, 'The time ball at Greenwich and the evolving methods of control', *Antiquarian Horology*, published as several parts in 2013: Part 1, 34, 2, pp.198-218; Part 2, 34, 3, pp.332-46; Part 3, 34, 4, pp.471-88
- Homes, Caitlin, 'The Astronomer Royal, the Hydrographer and the Time Ball: Collaborations in Time Signalling 1850-1910', *The British Journal for the History of Science*, 42, 3, 2009, pp.381-406

**George Biddell Airy, seventh Astronomer Royal**
- Airy, George Biddell, *Autobiography of Sir George Biddell Airy*, Wilfrid Airy (ed.), Cambridge University Press, 1896
- Perkins, A., 'Extraneous government business: the Astronomer Royal as government scientist: George Airy and his work on the commissions of state and other bodies, 1838-1880', *Journal of Astronomical History and Heritage*, 4, 2, 2001, pp.143-54
- ✦ Airy, George Biddell, *Autobiography of Sir George Biddell Airy*, Wilfrid Airy (ed.), Cambridge University Press, 1896, p.9

**40. Engraving of Cambridge Observatory**
- Airy, George Biddell, *Account of the Northumberland Equatorial and Dome, Attached to the Cambridge Observatory*, Cambridge University Press, 1844
- Airy, George Biddell, *Autobiography of Sir George Biddell Airy*, Wilfrid Airy (ed.), Cambridge University Press, 1896, chapter 4

- Hutchins, Roger, *British University Observatories, 1772–1939*, Ashgate Publishing, Ltd., Aldershot, 2008, sections 1.6 and 1.8
- Phillips, C., 'Robert Woodhouse and the Evolution of Cambridge Mathematics', *History of Science*, 44, 2006, pp.69–93
- Wilkes, M.V., 'Herschel, Peacock, Babbage and the Development of the Cambridge Curriculum', *Notes and Records of the Royal Society of London*, 44, 2, 1990, pp.205–19

### 41. Sketch of Playford cottage
- Airy, George Biddell, *Autobiography of Sir George Biddell Airy*, Wilfrid Airy (ed.), Cambridge University Press, Cambridge, 1896, pp.18–21, 182–3
- Suffolk Artists, 'Anna Airy, 1882–1964', https://suffolkartists.co.uk/index.cgi?choice=painter&pid=18 (accessed 27 December 2024)
- ✣ Airy, George Biddell, *Autobiography of Sir George Biddell Airy*, Wilfrid Airy (ed.), Cambridge University Press, 1896, p.57

### 42. Airy's hole punch
- Airy, George Biddell, *Autobiography of Sir George Biddell Airy*, Wilfrid Airy (ed.), Cambridge University Press, Cambridge, 1896, pp.72, 85, 158
- Belteki, Daniel, 'The winter of raw computers: the history of the lunar and planetary reductions of the Royal Observatory, Greenwich', *British Journal for the History of Science* 56, 1, 2023, pp.65–81
- Chapman, Allan, 'Airy's Greenwich Staff', *The Antiquarian Astronomer*, 6, 2012, pp.4–18
- Glass, Ian, 'The SAAO Astronomical Museum Observatory, Cape Town', p.9 https://www.saao.ac.za/~isg/museum.pdf (accessed 27 December 2024)
- RGO Archives, 'Correspondence with tradesmen, 1852', RGO 6/725, letter from Airy to Ransome & May outlining his design for the punch, 29 Sept 1852, f.687
- RGO Archives, 'Correspondence with tradesmen, 1854, RGO 6/728, letter from Airy to Ransome & May outlining his design for the punch, 12 September 1854, section 13, f.162
- Smith, Robert 'A National Observatory Transformed: Greenwich in the Nineteenth Century', *Journal for the History of Astronomy*, 22, 1991, pp.5–20
- ✣ Edwin Dunkin, 'A Day at the Observatory', *The Leisure Hour*, 524, 9 January 1862, pp.22–6
- ✣ 'No paper whatsoever to be destroyed', RGO Archives, 'General Order', 6 December 1860, in 'Occasional orders to Assistants, 1860–1869', RGO 6/40, f.10r

### 43. Grand orrery
- King, Henry C. and Millburn, John R., *Geared to the Stars: The Evolution of Planetariums, Orreries and Astronomical Clocks*, University of Toronto Press, 1978
- Sheehan, William, Bell, Trudy E., Kennett, Carolyn and Smith, Robert W., *Neptune: From Grand Discovery to a World Revealed; Essays on the 200th Anniversary of the Birth of John Couch Adams*, Springer Cham, 2021

### 44. Regulator no.587 by Dent
- Airy, George Biddell, 'VI. On the invention of a clock escapement', *Monthly Notices of the Royal Astronomical Society*, 5, 25, 1842, pp.221–2
- Anon., 'Horological Instruments in the Great Exhibition', *The Illustrated Exhibitor, a tribute to the world's industrial jubilee; comprising sketches, by pen and pencil, of the principal objects in the Great Exhibition of the Industry of All Nations*, no.23, 8 November 1851, pp.416–22; regulator shown on p.422
- Bennett, J. A., 'Airy and horology', *Annals of science*, 27, 1980, pp.269–85.
- Gillin, Edward J., 'Mechanics and mathematicians: George Biddell Airy and the social tensions in constructing time at Parliament, 1845–1860', *History of Science*, 58, 3, 2020, pp.301–25
- Mercer, Vaughan, *The life and letters of Edward John Dent, chronometer maker, and some account of his successors*, The Antiquarian Horological Society, London, 1977, pp.121–5, 161–5
- ✣ RGO Archives, 'Observations, correspondence and other papers, 1812–1831', copy of a letter to R. Byam (Office of Ordnance) concerning two chronometers by Dent, 28 March 1829, RGO 5/233, f.48r

### 45. Daylight View of the Great Comet of 1843
- Belteki, Daniel, '"There dwells some wish in every heart": the friendship between James David Forbes and Sir George Biddell Airy', University of St Andrews collections blog, https://university-collections.wp.st-andrews.ac.uk/2019/11/13/there-dwells-some-wish-in-every-heart-the-friendship-between-james-david-forbes-and-sir-george-biddell-airy/ (accessed 27 January 2025)
- Brück, Hermann A. and Brück, Mary T., *The Peripatetic Astronomer: The Life of Charles Piazzi Smyth*, Hilger, Bristol, 1988
- Stott, Carole and Hughes, David, 'Two Piazzi Smyth Comet Paintings', *Annals of Science*, 46, 1989, pp.165–72
- Warner, Brian, *Charles Piazzi Smyth: astronomer-artist: his Cape years, 1835–1845*, University of Cape Town Press, 1983

### The Age of Magnetism
- Cook, Alan, 'Edmond Halley and the Magnetic Field of the Earth', *Notes and Records of the Royal Society of London*, 55, 3, 2001, pp.473–90
- Macdonald, Lee, 'The origins and early years of the Magnetic and Meteorological department at Greenwich Observatory, 1834-1848', *Annals of Science*, 75, 3, 2018, pp.201–33
- Malin, Stuart, 'Geomagnetism at the Royal Observatory, Greenwich', *Quarterly Journal of the Royal Astronomical Society*, 37, 1996, pp.65–74
- ✣ John Barrow, *An auto-biographical memoir of Sir John Barrow, bart., late of the Admiralty, including reflections, observations and reminiscences, at home and abroad. From early life to advanced age*, John Murray, London, 1847, p.333

### 46. The Magnet House
- Airy, George Biddell, *Annual Report to the Board of Visitors*, 5 June 1841, p.1
- Glaisher, James, 'The Magnetic and Meteorological Department, just completed at Greenwich', *Illustrated London News*, 16 March 1844, pp.163–4
- Macdonald, Lee, 'The origins and early years of the Magnetic and Meteorological department at Greenwich Observatory, 1834-1848', *Annals of Science*, 75, 3, 2018, pp.201–33

### 47. Airy Transit Circle
- Airy, George Biddell, 'Appendix I. Description of the Transit Circle of the Royal Observatory, Greenwich', *Greenwich Observations in Astronomy, Magnetism and Meteorology made at the Royal Observatory*, Series 2, 14, 1854, pp.1–23 and plates I–XVI
- Belteki, Daniel, 'The Galvanic Connections of the Airy Transit Circle and the Airy Chronograph', *Bulletin of the Scientific Instrument Society*, 149, 2021, pp.2–9
- Gillin, Edward J., 'Tremoring transits: railways, the Royal Observatory and the capitalist challenge to Victorian astronomical science', *British Journal for the History of Science*, 53, 1, 2002, pp.1–24
- Howse, Derek, *Greenwich Observatory: The Royal Observatory at Greenwich and Herstmonceux 1675-1975, Volume 3: The Buildings and Instruments*, London: Taylor and Francis, 1975, pp.58–60, 109–10
- Satterthwaite, Gilbert E., 'Airy's transit circle', *Journal of Astronomical History and Heritage*, 4, 2, 2001, pp.115–41

### 48. The fireball meteor of 1850
- Anon., 'Splendid Meteor, Seen on Monday Night', *Illustrated London News*, 16 February 1850, pp.118–20
- Olson, Roberta J. and Pasachoff, Jay M., *Fire in the Sky: Comets and Meteors, the Decisive Centuries, in British*

- *Art and Science*, Cambridge University Press, 1999
- ✢ Glaisher, J., 'XXVIII. On the Meteor which appeared on Monday, the 11th of February 1850, at about 10h45m p.m.', *The London, Edinburgh and Dublin Philosophical Magazine and Journal of Science*, Series 3, 36, January–June 1850, pp.221–34
- ✢ Glaisher, J., 'XXXII. Additional Observations on the Meteor of February 11th, 1850, and Deductions of the Results from all the Observations, *The London, Edinburgh and Dublin Philosophical Magazine and Journal of Science*, Series 3, 36, January–June 1850, pp.249–71

### 49. Shepherd Motor Clock and Gate Clock
- Howse, Derek, *Greenwich Time and the Longitude*, Philip Wilson, London, 1997, pp.95–115

### 50. British local time map
- Airy, George Biddell, 'Account of Observations of the Total Solar Eclipse of 1860, July 18', *Monthly Notices of the Royal Astronomical Society*, 21, 1, 1860, pp.1–16
- Anon., 'Obituary: Henry Samuel Ellis', *Monthly Notices of the Royal Astronomical Society*, 39, 4, 1879, pp.225–6
- Grace's Guide, 'Henry Samuel Ellis (1825–1878)', https://www.gracesguide.co.uk/Henry_Samuel_Ellis (accessed 28 December 2024)
- Lyman, Ian P., *Railway Clocks*, Mayfield Books, Ashbourne, 2004, pp.21–44
- RGO Archives, 'Miscellaneous papers on chronometers, 1849–1852', RGO 6/597, maps at f.147 and f.149 with a related letter at f.146

### 51. Marine chronometer no.10 by Barraud
- Akkermans, Emily, 'Chronometers and chronometry on British Voyages of Exploration, 1819–1836', PhD diss., University of Edinburgh, 2021
- Betts, Jonathan, *Marine Chronometers at Greenwich*, Oxford University Press, 2017, pp.243–6
- Ellis, William, 'Lecture on the treatment of chronometers at the Royal Observatory, Greenwich', *The Horological Journal*, VII, 1866, pp.85–92
- Gould, Rupert, *Marine Chronometer: Its history and development*, J.D. Potter, London, 1923, pp.253–64
- Airy, George Biddell, *Annual Report to the Board of Visitors*, 1 June 1839, p.6
- ✢ Maunder, E. Walter, *The Royal Observatory: A Glance at its History and Work*, The Religious Tract Society, London, 1900, p.169

### 52. Public imperial standards of length
- Airy, George Biddell, 'Account of the Construction of the New National Standard of Length, and of its Principal Copies', *The Philosophical Transactions of the Royal Society*, 147, 1858, pp.621–702 (Section IX)
- RGO Archives, 'Astronomer Royal's journal, 1848–1861', RGO 6/25, f.172r
- Rooney, David, 'Rooney Vision: Public standards of length', https://rooneyvision.substack.com/p/public-standards-of-length (accessed 31 January 2025)
- Simpson, A.D.C., 'The pendulum as the British length standard: a nineteenth century legal aberration', in *Making Instruments Count: Essays on Historical Scientific Instruments Presented to Gerard L'Estrange Turner*, Robert Anderson, James A. Bennett and William F. Ryan, Variorum, Aldershot, 1993, pp.174–90

### 53. Magnetogram of the Carrington Event, 1 September 1859
- Clark, Stuart, *The Sun Kings: The Unexpected Tragedy of Richard Carrington and the Tale of How Modern Astronomy Began*, Princeton University Press, 2009
- ✢ Carrington, Richard, 'Description of a Singular Appearance seen in the Sun on September 1, 1859', *Monthly Notices of the Royal Astronomical Society*, 20, 1, 1859, pp.13–15

### 54. Airy's dip circle
- Airy, George Biddell, *Annual Report to the Board of Visitors*, 6 June 1857, p.12
- Airy, George Biddell, 'Description of the New Instrument by Simms entitled Airy's Instrument', in *Results of the Magnetic and Meteorological Observations made the Royal Observatory Greenwich, 1864*, pp.iii–v
- ✢ Airy, George Biddell, *Annual Report to the Board of Visitors*, 4 June 1859, p.13

### 55. Hourly time signal relay
- Akkermans, Emily, 'Wired in time: the neglected infrastructure of the Greenwich Time System', *Science Museum Group Journal*, 20, 23, 2023, https://journal.sciencemuseum.ac.uk/article/revealing-observatory-networks-through-object-stories-instrumental-networks/#abstract (accessed 14 March 2025)
- Ellis, William, 'Description of the Greenwich Time-Signal System', *Greenwich Observations*, 1879, Appendix.
- Ishibashi, Y., 'Constructing the 'automatic' Greenwich time system: George Biddell Airy and the telegraphic distribution of time, c.1852–1880', *British Journal for the History of Science*, 53, 1, 2020, pp.25–46
- ✢ Airy, George Biddell, *Annual Report to the Board of Visitors*, 4 June 1853, p.8

### 56. Single-needle telegraph
- Alice Springs Telegraph Station, https://alicespringstelegraphstation.com.au/ (accessed 31 January 2025)
- Christie, William, *Results of the Magnetical and Meteorological Observations made at The Royal Observatory, Greenwich, in the year 1882*, HMSO, London, 1884, p.xxxii
- Cryle, Denis, *Behind the Legend: The Many Worlds of Charles Todd*, Australian Scholarly Publishing, Melbourne, 2017
- Hallas, Sam, 'The Single Needle Telegraph', http://www.samhallas.co.uk/railway/single_needle.htm (accessed 31 December 2021)
- Howse, Derek, *Greenwich time and the discovery of the longitude*, Oxford University Press, 1980, pp.89–103
- Perry, C. R., 'The Rise and Fall of Government Telegraphy in Britain', *Business and Economic History*, 26, 2, 1997, pp.416–25

### Meteorology becomes a science
- Anderson, Katharine, *Predicting the Weather: Victorians and the Science of Meteorology*, University of Chicago Press, 2005, p.92
- Glaisher, James, *Report on the meteorology of London, and its relation to the epidemic of cholera*, HMSO, London, 1855
- Hunt, J., 'The handlers of time: The Belville Family and the Royal Observatory, 1811–1939', *Astronomy & Geophysics*, 40, 1, 1999, 1.23–1.27
- Macdonald, Lee, 'The origins and early years of the Magnetic and Meteorological department at Greenwich Observatory, 1834–1848', *Annals of Science*, 75, 3, 2018, pp.201–33
- Met Office, 'National Meteorological Library and Archive Factsheet 21 – Met Office History and Timeline', https://www.metoffice.gov.uk/binaries/content/assets/metofficegovuk/pdf/research/library-and-archive/library/publications/factsheets/factsheet_21-met-office-history-and-timeline_2024.pdf (accessed 31 January 2025)
- ✢ Marriott, William, 'Some account of the meteorological work of the late James Glaisher, FRS', *Quarterly Journal of the Royal Meteorological Society*, XXX, 129, 1904, pp.1–28,
- ✢ Glaisher, James, 'Report upon the meteorology of India in relation to the health of the troops there stationed', in *Volume I: Report of the Commissioners, Royal Commission on the Sanitary State of the Army in India*, HMSO, London, 1863
- ✢ RGO Archives, 'Papers on meteorology, 1861–1865', RGO 6/703, letter from George Biddell Airy to Robert Main, 20 February 1861, in Section 2, f.44

**57. Public barometer**
- A London Inheritance, 'Negretti & Zambra, Admiral FitzRoy, James Glaisher. From London to Orkney via Greenwich', https://alondoninheritance.com/london-characters/negretti-zambra-admiral-fitzroy-james-glaisher-from-london-to-orkney-via-greenwich/ (accessed 8 February 2025)
- Anon., 'Barometers for Life-Boat Stations', *The Life-Boat*, 4, 38, 1 October 1860, pp.336–9
- Glaisher, J., 'On the Northumberland Coast Station Barometers', *The Life-Boat*, 4, 39, 1 January 1861, pp.367–71
- Negretti, Henry and Zambra, Joseph Warren, *A Treatise on Meteorological Instruments*, Negretti and Zambra, London, 1864, p.12
- Spencer Jones, Harold, *Annual Report to the Board of Visitors*, 1 June 1946, p.1
✢ Airy, George Biddell, *Annual Report to the Board of Visitors*, 3 June 1865, p.19

**58. Earth current galvanometer**
- Airy, George Biddell, 'Comparison of magnetic disturbances recorded by self-registering magnetometers at the Royal Observatory, Greenwich, with magnetic disturbances described from the corresponding terrestrial galvanic currents recorded by the self-registering galvanometers of the Royal Observatory', *Philosophical Transactions of the Royal Society*, 158A, 1868, pp.465–72
✢ Airy, George Biddell, *Annual Report to the Board of Visitors*, 7 June 1862, p.4
✢ Christie, William, *Results of the Magnetical and Meteorological Observations made at The Royal Observatory, Greenwich, in the year 1882*, HMSO, London, 1884, p.xxxii

**59. The Airy family**
- Airy, George Biddell, *Autobiography of Sir George Biddell Airy*, Wilfrid Airy (ed.), Cambridge University Press, 1896
- RGO Archives, 'First Assistant's correspondence, 1849–1854', RGO 6/29, letter from Robert Main to Miss Airy, 15 August 1850
- Frances Ward, 'The Airys at Greenwich', *Journal of Astronomical History and Heritage*, 4, 2, 2001, pp.155–61

**60. The Midnight Sky (1869)**
- Lechner, Doris, *Histories for the Many: The Victorian Family Magazine and Popular Representations of the Past. The Leisure Hour, 1852-1870*, Transcript Verlag, Bielefeld, 2016
✢ 'Literature', *Illustrated London News*, 4 December 1869, p.559

**Cable production in Greenwich**
- Alcatel Submarine Networks, https://www.asn.com/ (accessed 8 February 2025)
- Arscott, Caroline and Pettitt, Clare, *Victorians Decoded: Art and Telegraphy*, The Courtauld Institute of Art, London, https://scrambledmessages.ac.uk/documents/52/Victorians_Decoded_P32Plrd.pdf (accessed 17 March 2025)
- Ash, Stewart, 'The Story of Subsea Telecommunications & its Association with Enderby House', https://atlantic-cable.com/CableCos/EnderbysWharf/Enderby_Telcoms_Story.pdf (accessed 14 March 2025)
- Bright, Charles, *Submarine Telegraphs: Their History, Construction and Working*, C. Lockwood and Son, London, 1898
- Burns, Bill, 'History of the Atlantic Cable & Undersea Communications from the first submarine cable of 1850 to the worldwide fiber optic network', https://atlantic-cable.com/ (accessed 26 March 2025)
- Ellis, William, 'Description of the Greenwich Time-Signal System', *Greenwich Observations*, 41, 1881, J1–J13
- Hunt, Bruce J., *Imperial Science: cable telegraphy and electrical physics in the Victorian British Empire*, Cambridge University Press, 2020
- Green, Francis Matthews, 'Telegraphic Determination of Longitude', *Popular Science Monthly*, 7, August 1875, pp.426–33
- Russell, W.H., *The Atlantic Telegraph*, Day and Son Ltd, London, 1865
✢ Bright, Charles, *The Story of the Atlantic Cable*, D. Appleton and Company, New York, 1903, p.35

**61. Presentation box of submarine cable samples**
- Airy, George Biddell, *Account of observations of the Transit of Venus, 1874, December 8*, HMSO, London, 1881
- Bright, Charles, *Submarine Telegraphs: Their History, Construction and Working*, C. Lockwood and Son, London, 1898
- Glover, Bill, 'The British-Indian Submarine Telegraph Company', https://atlantic-cable.com/Cables/1870BritishIndian/ (accessed 8 February 2025)
- Headrick, Daniel R. and Griset, Pascal, 'Submarine Telegraph Cables: Business and Politics, 1838-1939', *The Business History Review*, 75, 3, 2001, pp.543–78

**62. Airy's Sèvres vase**
- Bureau international des poids et mesures (BIPM), 'The Metre Convention', https://www.bipm.org/en/metre-convention (accessed 8 February 2025)
- Dyson, Frank Watson, *Annual Report of the Astronomer Royal*, 6 June 1925, p.5
- Perkins, Adam, '"Extraneous government business": the Astronomer Royal as government scientist: George Airy and his work on the commissions of state and other bodies, 1838-1880', *Journal of Astronomical History and Heritage*, 4, 2, 2001, pp.143–54
- Quinn, Terry and Kovalevsky, Jean, 'The development of modern metrology and its role today', *Philosophical Transactions of the Royal Society A*, 363, 2005, pp.2, 307-27
- RGO Archives, 'Correspondence on standards, 1873–1876', RGO 6/367

**63. Imperial Order of the Rose**
- RGO Archives, 'Observatory Journal', entry for 17 July 1871, RGO 6/26
- Barman, Roderick J., *Citizen Emperor: Pedro II and the Making of Brazil, 1825–1891*, Stanford University Press, 1999, pp.117–18, 235–7
- Airy, George Biddell, *Annual Report to the Board of Visitors*, 1 June 1872, p.22
✢ Airy, George Biddell, *Autobiography of Sir George Biddell Airy*, Wilfrid Airy (ed.), Cambridge University Press, 1896, p.299

**64. Dallmeyer photoheliograph**
- Dolan, Graham, 'Telescope: The five Dallmeyer Photoheliographs (1873)', http://www.royalobservatorygreenwich.org/articles.php?article=1251 (accessed 14 March 2025)
- Maunder, E. Walter, *The Royal Observatory, Greenwich. A Glance at Its History and Work*, The Religious Tract Society, London, 1900, chapter 10
- Airy, George Biddell, *British expeditions for the observation of Venus, 1874, December 8. Instructions to observers*, HMSO, London, 1874

**65. Test plate for the Janssen photographic revolver**
- Débarbat, S. and Launay, F., 'The 1874 Transit of Venus Observed in Japan by the French, and Associated Relics', *Journal of Astronomical History and Heritage*, 9, 2, 2006, pp.167–71
- Launay, Françoise and Hingley, Peter D., 'Jules Janssen's "Revolver Photographique" and its British derivative, "The Janssen Slide"', *Journal for the History of Astronomy*, 36, 1, 122, 2005, pp.57–79
- Ratcliff, Jessica, 'Models, metaphors, and the Transit of Venus in Victorian Britain', *Cahiers François Viète*, 11-12, 2006, pp.63–82

**66. Sunshine recorder**
- Airy, George Biddell, *Annual Report to the Board of Visitors*, 7 June 1879, p.17
- Campbell, J. F., 'Campbell's Registering Sun-Dial', *Good Words for 1879*, 20, pp.706–13
- Christie, William, *Annual Report to the Board of Visitors*, 4 June 1887, p.15
- Christie, William, *Annual Report to the Board of Visitors*, 6 June 1896, p.15

- Ellis, William, 'XLIV. Results derived from the Sunshine Records obtained at the Royal Observatory, Greenwich, by means of Campbell's Self-Registering Sun Dial, during the year ending April 30th, 1877', *Quarterly Journal of the Royal Meteorological Society*, 3, 24, 1877, pp.460-7
- Sanchez-Lorenzo, A., Calbó, J., Wild, M., Azorin-Molina, C. and Sanchez-Romero, A., 'New insights into the history of the Campbell-Stokes sunshine recorder', *Weather*, 68, 2013, pp.327-31
✢ Spencer Jones, Harold, *Annual Report to the Board of Visitors,* 3 June 1939, p.43

**Splitting starlight with spectroscopy**
- Becker, B.J., *Unravelling Starlight: William and Margaret Huggins and the Rise of the New Astronomy*, Cambridge University Press, p.336
- Dewhirst, David and Hoskins, Michael, 'Chapter 8: The message of starlight: the rise of astrophysics', in *The Cambridge Concise History of Astronomy*, Michael Hoskins (ed.), Cambridge University Press, 1999, pp.224-7
- Hirshfeld, Alan, *Starlight Detectives: How Astronomers, Inventors, and Eccentrics Discovered the Modern Universe*, Bellevue Literary Press, New York, 2014
- Newton, Isaac, 'A Letter of Mr. Isaac Newton ... containing his New Theory about Light and Colors', *Philosophical Transactions of the Royal Society*, 80, 19 Feb. 1671/2, pp.3,075-87
✢ Kirchhoff, Gustav and Bunsen, Robert, 'Chemical Analysis by Observations of Spectra', *Annalen der Physik und der Chemie*, 110, 1860, pp.161-89. English translation available at https://www.chemteam.info/Chem-History/Kirchhoff-Bunsen-1860.html (accessed 16 March 2025)
✢ RGO Archives, 'Spectroscopic Observations', RGO 6/759, entry for 29 July 1874
✢ Greaves, W.M.H., Davidson, C. and Martin, E., 'The zero point of Greenwich colour temperature system', *Monthly Notices of the Royal Astronomical Society*, 94, 507, 1934, pp.488-507

**67. Two-prism spectroscope**
- Airy, George Biddell, 'Spectroscopic Observations Made at the Royal Observatory, Greenwich, in the Years 1874 and 1875', *Greenwich Observations in Astronomy, Magnetism and Meteorology made at the Royal Observatory*, 2, 37, 1877, pp.99-130
- Becker, B.J., *Unravelling Starlight: William and Margaret Huggins and the Rise of the New Astronomy*, Cambridge University Press, pp.222-3, 311-12
- Maunder, E. Walter, *The Royal Observatory: A Glance at its History and Work*, The Religious Tract Society, London, 1900
- RGO Archives, 'Spectroscopic observations, 1874-1889', RGO 6/759
✢ Maunder, E. Walter, T*he Royal Observatory: A Glance at its History and Work*, The Religious Tract Society, London, 1900, p.274

**68. Kew-pattern unifilar magnetometer**
- Airy, George Biddell, 'Observations for the Absolute Measure of the Horizontal Force of Terrestrial Magnetism', *Results of the Magnetic and Meteorological Observations*, 1862, pp.xxxvii-xxxix
- Bryden, David, 'Quality Control in the Making of Scientific Instruments: Kew Observatory and the Verification of Meteorological, Magnetic and Other Instruments, 1851-1899', *Bulletin of the Scientific Instrument Society*, 88, 2006, pp.48-59
- W.W., 'Magnetometer', *Encyclopaedia Britannica*, 11th edition, 17, 1911, pp.386-8

**William Henry Maloney Christie, eighth Astronomer Royal**
- Airy, George Biddell, 'Spectroscopic Observations at the Royal Observatory', *Greenwich Observations*, 1874, 1875, 1877 and 1879
- Dolan, Graham, 'People: William Christie, Astronomer Royal', http://www.royalobservatorygreenwich.org/articles.php?article=1165 (accessed 4 April 2024)
- Meadows, A., 'Christie, Sir William Henry Mahoney (1845-1922), astronomer', *ODNB*, https://www.oxforddnb.com/display/10.1093/ref:odnb/9780198614128.001.0001/odnb-9780198614128-e-32409 (accessed 4 April 2024)

**69. The 1884 International Meridian Conference**
- Howse, Derek, *Greenwich Time and Longitude*, Philip Wilson, London, 1997, pp.125-43
- Project Gutenberg, 'International Conference Held at Washington for the Purpose of Fixing a Prime Meridian and a Universal Day' [full transcript of the conference proceedings], https://www.gutenberg.org/ebooks/17759 (accessed 14 March 2025)
- Withers, Charles W., *Zero Degrees: Geographies of the Prime Meridian*, Harvard University Press, Cambridge, MA, 2017, chapters 5 and 6
✢ Project Gutenberg, 'International Conference Held at Washington for the Purpose of Fixing a Prime Meridian and a Universal Day' [full transcript of the conference proceedings], https://www.gutenberg.org/ebooks/17759 (accessed 14 March 2025)

**70. World time converter**
- Howse, Derek, *Greenwich Time and the Longitude*, Philip Wilson, London, 1997
- Ogle, Vanessa, *The Global Transformation of Time: 1870-1950*, Harvard University Press, 2015
✢ Project Gutenberg, 'International Conference Held at Washington for the Purpose of Fixing a Prime Meridian and a Universal Day', [full transcript of the conference proceedings], https://www.gutenberg.org/ebooks/17759 (accessed 14 March 2025)

**71. Astrographic telescope**
- Chinnici, Ileana, 'Precursors to IAU: Paris Observatory and the Carte du Ciel Project', in *Astronomers as Diplomats, Historical & Cultural Astronomy*, T. Montmerle and D. Fauque (eds), Springer Nature Switzerland, 2022
- Christie, W., *Astrographic Catalogue 1900.00, Greenwich Section, Dec. +64° to 90°, Volume I*, HMSO, Edinburgh, 1904
- Haley, Paul, 'Entente céleste: David Gill, Ernest Mouchez, and the Cape and Paris Observatories 1878-92', *Antiquarian Astronomer*, 10, 2016, pp.13-37
- Jones, Derek, 'The scientific value of the Carte du Ciel', *Astronomy and Geophysics*, 41, 2000, 5.16-5.20

**The women of the Carte du Ciel**
- Bigg, Charlotte, 'Photography and labour history of astrometry: The Carte du Ciel', in *The Role of Visual Representations in Astronomy: History and Research Practice* (Acta Historica Astronomiae series, volume 9), Klaus Hentschel and Axel D. Wittmann (eds), Verlag Harri Deutsch, Leipzig, 2000, pp.90-106
- Bodleian Library, Oxford, RGO Plate Collection, crate 80 (Astrographic Plates) and crate 105 (observing notebooks).
- Corral, Moreno, Arturo, Marco and Schuster, William J., 'The Mexican astrographic catalogue and Carte du Ciel Project', *Journal of Astronomical History and Heritage*, 23, 3, 2020, pp.601-13
- Dolan, Graham, 'Christie's "Lady Computers" - the astrographic pioneers of Greenwich', http://www.royalobservatorygreenwich.org/articles.php?article=1280 (accessed 19 February 2025)
✢ Anon., 'A student of the stars: half an hour with Miss Alice Everett, M.A.', *The Sketch*, 22 November 1893, p.192

**72. Solar plate micrometer**
- Dolan, Graham, 'Telescope: The five Dallmeyer Photoheliographs (1873)', http://www.royalobservatorygreenwich.org/articles.php?article=1251 (accessed 27 February 2025)

- Dolan, Graham, 'People: Annie Scott Dill Maunder (née Russell)', http://www.royalobservatorygreenwich.org/articles.php?article=1274 (accessed 27 February 2025)
- Willis, D.M., Wild, M.N., Appleby, G.M. and Macdonald, L.T., 'The Greenwich Photo-heliographic Results (1874–1885): Observing Telescopes, Photographic Processes, and Solar Images', *Solar Physics*, 291, 2016, pp.2, 553–86

**73. 'Greetings from Greenwich' postcard**
- Johnston, S., 'Managing the observatory: discipline, order and disorder at Greenwich, 1835–1933', *British Journal for the History of Science*, 54, 2, 2021, pp.155–75
- Laurie, P.S., 'The Board of Visitors of the Royal Observatory-II: 1830–1965', *Quarterly Journal of the Royal Astronomical Society*, 8, 1967, pp.334–53
- Wilson, Margaret, *Ninth Astronomer Royal: The life of Frank Watson Dyson*, W. Heffer and Sons Ltd, Cambridge, 1951, pp.176–7
✢ Astor, P., 'The Centre of the World – How Time is Made', *The Harmsworth Magazine*, 6, 1901, pp.247–52

**74. Pocket chronometer 'Arnold 485'**
- Hunt, J., 'The handlers of time: The Belville Family and the Royal Observatory, 1811–1939', *Astronomy & Geophysics*, 40, 1, 1999, 1.23–1.27
- Rooney, David, *Ruth Belville: The Greenwich Time Lady*, National Maritime Museum, London, 2008

**75. 28in. Great Equatorial Telescope**
- Christie, W., 'On a New Dome to be Erected at the Royal Observatory, Greenwich', *Monthly Notices of the Royal Astronomical Society*, 51, 7, 1891, pp.436–8
- Glass, I., *Victorian telescope makers: the lives and letters of Thomas and Howard Grubb*, Institute of Physics Publishing, Bristol, 1997
- Wright, D., 'The 28-inch Refractor at Greenwich – a History of Two Telescopes', *Quarterly Journal of the Royal Astronomical Society*, 31, 4, 1990, pp.551–6
✢ National Archives, Kew, 'Minutes of the Board of Visitors, 1885, Royal Observatory', Records of the Admiralty, ADM 190/6/8

**76. *The Secret Agent* (1907)**
- Higgitt, Rebekah, 'The real story of the Secret Agent and the Greenwich Observatory bombing', *The Guardian*, 5 August 2016, https://www.theguardian.com/science/the-h-word/2016/aug/05/secret-agent-greenwich-observatory-bombing-of-1894 (accessed 10 June 2024)
- RGO Archives, 'Astronomer Royal's journal, 1893–1898', RGO 7/30 f.15
- Rooney, David, *Ruth Belville: The Greenwich Time Lady*, National Maritime Museum, London, 2008, pp.66–70
- Royal Observatory Edinburgh, 'Bomb attack at the Royal Observatory Edinburgh', https://www.roe.ac.uk/whatsnew/event/20130521/ (accessed 31 May 2024)
- Wilson, Margaret, *Ninth Astronomer Royal: The life of Frank Watson Dyson*, W. Heffer and Sons Ltd, Cambridge, 1951, p.163
✢ RGO Archives, 'Papers on Greenwich Park, 1876–1912', RGO 7/58, f.1

**77. Altazimuth Pavilion**
- Dolan, Graham, 'The Altazimuth Pavilion', http://www.royalobservatorygreenwich.org/articles.php?article=918 (accessed 17 July 2024)
- Dolan, Graham, 'Telescope: Christie's New Altazimuth Telescope (1896)', http://www.royalobservatorygreenwich.org/articles.php?article=1090 (accessed 17 July 2024)
- Ridpath, Ian, 'A brief history of Halley's Comet: The return of 1910', http://www.ianridpath.com/halley/halley10.html (accessed 10 June 2024)

**78. New Physical Observatory**
- Dolan, Graham, 'The South Building (formerly the New Physical Observatory), the Lassell Dome, & the Lower Store', http://www.royalobservatorygreenwich.org/articles.php?article=920 (accessed 17 July 2024)
- Higgitt, Rebekah, 'A British national observatory: the building of the New Physical Observatory at Greenwich, 1889–1898', *British Journal for the History of Science*, FirstView, 2013, pp.1–27

**Sports and social clubs**
- Dominici, Sara, 'Early photographic federations and the pursuit of collaborative education', *Early Popular Visual Culture*, 20, 4, 2022, pp.388–411
- Holt, Richard, *Sport and the British: A Modern History*, Oxford University Press, 1989, pp.84–6, 96–7
- RGO Archives, 'Miscellaneous correspondence, 1872–1920', RGO 7/28
- RGO Archives, 'Chief Assistant's journal, 1879–1892', RGO 7/29
- RGO Archives, 'Papers on societies, 1877–1910', RGO 7/222
- Wilkins, George A., 'A Personal History of the Royal Greenwich Observatory at Herstmonceux Castle, 1948–1990; Appendix D: The Royal Greenwich Observatory Club at Herstmonceux Castle', on deposit at the RGO Archives and available online: https://www.lib.cam.ac.uk/collections/departments/manuscripts-university-archives/significant-archival-collections/royal-0 (accessed 29 December 2024)
- Witchell, W. M., 'RO Hockey Club', *The Castle Review: The Journal of the Royal Observatory Sports and Social Club*, published across several issues from April to November 1951. Abridged version available at Graham Dolan, 'Royal Observatory Hockey Club (ROHC)', http://www.royalobservatorygreenwich.org/articles.php?article=1198 (accessed 28 December 2024)
✢ Witchell, W.M., 'RO Hockey Club', *The Castle Review: The Journal of the Royal Observatory Sports and Social Club*, published across several issues from April to November 1951. Abridged version available at Graham Dolan, 'Royal Observatory Hockey Club (ROHC)', http://www.royalobservatorygreenwich.org/articles.php?article=1198 (accessed 28 December 2024)
✢ RGO Archives, 'Miscellaneous correspondence, 1872–1920', RGO 7/28, Sixth Exhibition catalogue, 1906

**79. Spider fork**
- Hunt, Frederick K., 'The planet watchers of Greenwich', *Household Words*, 1, 9, 25 May 1850, pp.200–04
- Hunt, Frederick K., 'Greenwich Weather Wisdom', *Household Words*, 1, 10, 1 June 1850, pp.222–5
- Maunder, E. Walter, 'Making a Spider Line Reticule', *Journal of the British Astronomical Association*, 4, 1894, pp.243–5
- RGO Archives, 'A catalogue of instruments and furniture belonging to the Royal Observatory, dated 4 June 1864, with lists of additions to 1875', RGO 6/63
- Turner, Steven, 'Spiders in the Crosshairs: Cobwebs, Instrument Makers, and the Search for the Perfect Line', *Journal of the Antique Telescope Society*, 1, 1992, pp.10–12
✢ RGO Archives, 'Papers on moveable property, 1873 –1875', RGO 6/63
✢ Knight Hunt, Frederick, 'The planet watchers of Greenwich', *Household Words*, 1, 9, 25 May 1850, pp.200–04
✢ Knight Hunt, Frederick, 'Greenwich Weather Wisdom', *Household Words*, 1, 10, 1 June 1850, pp.222–5

**80. Article by Maunder about the 'canals' on Mars**
- Anon., 'Physical Observations of Mars, made at the Royal Observatory, Greenwich', *Monthly Notices of the Royal Astronomical Society*, 38, 1877, pp.34–8
- Burnett, John, 'British studies of

- Mars: 1877-1914', *Journal of the British Astronomical Association*, 89, 2, 1979, pp.136–43
- Crowe, Michael, *The Extraterrestrial Life Debate 1750-1900: The Idea of a Plurality of Worlds from Kant to Lowell*, Cambridge University Press, 1986
- Lane, K. Maria, *Geographies of Mars: Seeing and Knowing the Red Planet*, University of Chicago Press, 2010
- Maunder, E. Walter, 'The Canals on Mars', *The Observatory*, 11, 1888, pp.345–8
- Maunder, E. Walter and Evans, J., 'Experiments as to the Actuality of the "Canals" observed on Mars', *Monthly Notices of the Royal Astronomical Society*, 63, 1903, pp.488–99
- Sheehan, W., *The Planet Mars: A History of Observation and Discovery*, The University of Arizona Press, Tucson, 1996
- ✢ Maunder, E. Walter, 'The Canals on Mars', *The Observatory*, 11, 1888, pp.345–8

## 81. The Heavens and their Story (1908)
- Hollis, H., 'Obituary: E.W. Maunder', *The English and Amateur Mechanics* 3, 487, 1928, p.507
- Kinder, A., 'Edward Walter Maunder FRAS (1851-1928): his life and times', *Journal of the British Astronomical Association*, 118, 1, 2008, pp.21–42
- Ogilvie, M., 'Obligatory amateurs: Annie Maunder (1868–1947) and British women astronomers at the dawn of professional astronomy', *British Journal for the History of Science*, 33, 1, 2000, pp.67–84
- ✢ Hollis, H., 'Mr E.W. Maunder – Nova Pictoris', *The English and Amateur Mechanics*, 3, 487, 13 April 1928, p.507

## 82. 30in. Thompson Reflector
- Central Bureau for Astronomical Telegrams (CBAT), 'IAUC 2846: N Mon 1975 (= A0620-00); N Cyg 1975; 1975h; 1975g; 1975i; Sats OF JUPITER' [naming of Jupiter VIII as 'Pasiphae'], http://www.cbat.eps.harvard.edu/iauc/02800/02846.html (accessed 16 March 2025)
- Christie, W. and Melotte, P., 'Note on the newly discovered Eighth Satellite of Jupiter, photographed at the Royal Observatory, Greenwich', *Monthly Notices of the Royal Astronomical Society*, 68, 6, 1908, pp.456–7
- Cowell, P. H. 'Note on the discovery of a moving object near Jupiter (1908 CJ), Greenwich, Royal Observatory', *Monthly Notices of the Royal Astronomical Society*, 68, 5, 1908, p.373
- Dolan, Graham, 'Telescope: The 26-inch Photographic Refractor & the 30-inch reflector of the Thompson Equatorial (1896)', http://www.royalobservatorygreenwich.org/articles.php?article=1093 (accessed 16 March 2025)
- Furner, H., 'Obituary: Philibert Jacques Melotte FRAS', *Journal of the British Astronomical Association*, 72, 1, 1962, pp.45–6
- Hunter, A., 'Obituary Notices: Philibert Jacques Melotte', *Quarterly Journal of the Royal Astronomical Society*, 3, 1962, pp.48–50

## Frank Watson Dyson, ninth Astronomer Royal
- Meadows, A., 'Dyson, Sir Frank Watson (1868–1939), astronomer', *ODNB*, https://www.oxforddnb.com/display/10.1093/ref:odnb/9780198614128.001.0001/odnb-9780198614128-e-32949 (accessed 4 April 2024)
- Spencer Jones, Harold, 'Obituary: Sir Frank Watson Dyson', *The Observatory*, 62, 1939, pp.179–87
- Wilson, Margaret, *Ninth Astronomer Royal: the life of Frank Watson Dyson*, W. Heffer and Sons, Ltd, Cambridge, 1951

## 83. Spherical plate calculator for star coordinates
- Anon., 'Obituary: Dr. W.B. Blaikie', *Nature*, 19 May 1928, p.801
- Dyson, Frank, 'The systematic motions of the Stars shown in the cross proper motions of the Bradley Stars', *Monthly Notices of the Royal Astronomical Society*, 70, 5, 1910, pp.416–29
- Kennefick, Daniel, *No shadow of a doubt: the 1919 eclipse that confirmed Einstein's Theory of Relativity*, Princeton University Press, 2019, pp.125–34
- Lenman, B., 'Blaikie, Walter Biggar (1847–1928)', *ODNB*, https://www.oxforddnb.com/display/10.1093/ref:odnb/9780198614128.001.0001/odnb-9780198614128-e-37198 (accessed 16 March 2025)
- Turner, H., 'On the Diagrammatic Representation of Proper Motions', *Monthly Notices of the Royal Astronomical Society*, 70, 3, 1910, pp.204–216

## 84. Henry Outhwaite, Observatory Secretary
- Dyson, Frank Watson, *Annual Report to the Board of Visitors*, 5 June 1915, p.21
- Dyson, Frank Watson, *Annual Report to the Board of Visitors*, 3 June 1916, p.15
- Melotte, P.J., 'Obituaries: H. Outhwaite', *The Observatory*, 70, 1950, p.32
- RGO Archives, 'Correspondence on military service, 1914–1919', RGO 8/36
- Wilson, Margaret, *Ninth Astronomer Royal: The life of Frank Watson Dyson*, W. Heffer and Sons Ltd, Cambridge, 1951, p.174
- ✢ RGO Archives, 'Correspondence on military service, 1914–1919', RGO 8/36, f.139, letter from Whitaker to Outhwaite, 2 December 1915

## 85. First World War binoculars
- Martin, E.G. 'Some historical items', *The Castle Review*, February 1952, available at RGO archives, '"The Castle Review", 1951-1963, 1965, 1981', RGO 80/9/1
- Reid, William, 'Binoculars and the National Service League', *Journal of the Society for Army Historical Research*, 84, 337, 2006, pp.42–51
- RGO Archives, 'Correspondence on the Observatory and Herstmonceux, 1934-1957', RGO 9/1
- Wilson, Margaret, *Ninth Astronomer Royal: The life of Frank Watson Dyson*, W. Heffer and Sons Ltd, Cambridge, 1951, p.175

## 86. Double star catalogue by Jonckheere
- Jonckheere, R., 'Measures of the Diameter of Mercury obtained at the Royal Observatory, Greenwich, during the Transit of 1914 November 6-7', *Monthly Notices of the Royal Astronomical Society*, 75, 1, 1914, pp.33–4
- Jonckheere, R., 'Lille Observatory and the war', *The Observatory*, 38, 485, 1915, pp.142–5
- Jonckheere, R., 'Catalogue and Measures of Double Stars Discovered Visually from 1905 to 1916 within 105° of the North Pole and Under 5" Separation', *Memoirs of the Royal Astronomical Society*, 61, 1917, pp.3–205
- ✢ Jonckheere, R., 'Catalogue and Measures of Double Stars Discovered Visually from 1905 to 1916 within 105° of the North Pole and Under 5° Separation', *Memoirs of the Royal Astronomical Society*, 61, 1917, pp.3–205

## 87. Glass photopositive of the 1919 total solar eclipse
- Anon., 'Court Circular', *The Times*, 11 June 1921, p.13
- Stanley, Matthew, '"An Expedition to Heal the Wounds of War": The 1919 Eclipse and Eddington as Quaker Adventurer', *Isis*, 94, 1, 2003, pp.57–89
- Wilson, Margaret, *Ninth Astronomer Royal: The life of Frank Watson Dyson*, W. Heffer and Sons Lt, Cambridge, 1951, pp.191–4

## 88. Measuring device for the star trail camera plates
- Anon., 'Royal Meteorological Society' [report of meeting held on 21 April 1920], *The Observatory*, 43, 552, 1920, pp.184–5
- Dyson, Frank Watson, *Results of the magnetic and meteorological observations made at the Royal Observatory Greenwich in the year 1920*, p.20

- Harlan, E. A., Walker, Merle F., 'Star-Trail Telescope for Astronomical Site-Testing', *Publications of the Astronomical Society of the Pacific*, 77, 457, 1965, p.246
- Kadel, B.C., 'An Improvement in the Pole Star Recorder', *Monthly Weather Review*, 1919, p.154
- Roth, A. Lawrence, *Sounding the Ocean of Air*, 1900, pp.22–3
✢ Spencer Jones, Harold, *Annual Report to the Board of Visitors*, 13 June 1953, p.21

### 89. Regulator no.2016 by Dent
- Akkermans, Emily, 'The history of the BBC pips', https://www.rmg.co.uk/stories/time/bbc-pips-royal-observatory-greenwich-time-service (accessed 8 March 2025)
- Dyson, Frank Watson, *Annual Report to the Board of Visitors*, 7 June 1924, p.15
- Miles, R.A., *Synchronome: Masters of Electrical Timekeeping*, Antiquarian Horological Society, Ticehurst, 2011, pp.204–07
- Todd, Mike, 'The Greenwich Time Signal', https://www.miketodd.net/other/gts.htm (accessed 2 June 2025)

### 90. Shortt free-pendulum clock system
- Anon., 'Obituary: William Hamilton Shortt, 1881–1971', *Proceedings of the Institution of Civil Engineers*, 50, 3, 1971, pp.396-7
- Hope-Jones, F., 'The Free Pendulum', *Journal of the Royal Society of Arts*, 72, 3731, 1924, pp.446–60
- Hope-Jones, F., *Electrical Timekeeping*, N.A.G. Press, 1949, pp.162–206
- Miles, R.A., *Synchronome: Masters of Electrical Timekeeping*, Antiquarian Horological Society, Ticehurst, 2011, pp.168–94
- Spencer Jones, Harold, *Annual Report to the Board of Visitors*, 5 June 1937, p.37

### 91. Glass photopositive of the Magnetic Pavilion at Abinger
- Dolan, Graham, 'Abinger Magnetic Observatory, 1923–1957', http://www.royalobservatorygreenwich.org/articles.php?article=911 (accessed 17 July 2024)

### 92. Observer's card for the Giggleswick total solar eclipse
- Marriott, R.A., '1927: a British eclipse', *Journal of the British Astronomical Association*, 109, 3, 1999, pp.117–43
- Wilson, Margaret, *Ninth Astronomer Royal: The life of Frank Watson Dyson*, W. Heffer and Sons Ltd, Cambridge, pp.224-7

### 93. Warren Synclock
- BBC News, 'Speaking Clock: Why are people still dialling for the time?', https://www.bbc.co.uk/news/magazine-14198506 (accessed 29 December 2024)
- Emmerson, Andrew, 'Speaking Clocks', https://www.britishtelephones.com/clocks/spkgclock.htm (accessed 29 December 2024)
- Edgcumbe, Everett, various advertisements for the 'Synclock', *The Horological Journal*, vol. LXXIII (February 1931), p.xiii; vol. LXXIII (May 1931), p.xiii; vol. LXXV (January 1933), p.vi; vol. LXXV (April 1933), front cover.
- Hope-Jones, F., *Electrical Timekeeping*, N.A.G. Press, London, 1940, pp.224–47
- Miles, R. A., *Synchronome: Masters of Electrical Timekeeping*, Antiquarian Horological Society, Ticehurst, 2011, pp.208–15
- Pike, William G., 'A brief description of TIM the G.P.O. Clock, which has answered over 20,000,000 calls since its installation just over a year ago', *Practical Mechanic*, May 1938, p.425
- Spencer Jones, Harold, 'The Royal Greenwich Observatory', *Proceedings of the Royal Society of London, Series B, Biological Sciences*, 136, 884 (Oct. 19, 1949), pp.349–77

### 94. Rupert Gould's notebooks
- Betts, Jonathan, *Time Restored: The Harrison Timekeepers and R.T. Gould, the Man who Knew (almost) Everything*, Oxford University Press, 2006
- Gould, Rupert, *Marine Chronometer: Its history and development*, J. D. Potter, London, 1923
- Gould, Rupert, 'John Harrison and his Timekeepers', *Mariner's Mirror*, XXI, 2, April 1935, pp.115–39
✢ Gould, Rupert, 'John Harrison and his Timekeepers', *Mariner's Mirror*, XXI, 2, April 1935, p.131

### 95. Mary French with the plate micrometer
- Christie, William H.M., 'A Micrometer for measuring the Plates of the Astrophotographic Chart', *Monthly Notices of the Royal Astronomical Society*, 53, 5, March 1893, pp.326–9
- Christie, William H.M., *Astrographic Catalogue 1900.00, Greenwich Section, Dec. +64° to 90°, Volume I*, HMSO, Edinburgh, 1904, pp.xix-xxii
- Dyson, Frank Watson, *Annual Report for the Board of Visitors*, 1 June 1929, p.7
- RGO Archives, 'Register of computers, 1836–1939', RGO 7/266
- RGO Archives, 'Royal Observatory, Greenwich and R.G.O. History Papers (including Philip S. Laurie Papers), 1761–c.1982', RGO 107
✢ Wilson, Margaret, *Ninth Astronomer Royal: The life of Frank Watson Dyson*, W. Heffer and Sons Ltd, Cambridge, p.186

### Harold Spencer Jones, tenth Astronomer Royal
- Woolley, Richard van der Riet, 'Obituary: Harold Spencer Jones, 1890-1960', *Biographical Memoirs of the Royal Society*, 7, 1961, pp.136–45

### 96. Slitless spectrograph
- Anon., 'The opening of the Yapp telescope at Greenwich', *The Observatory*, 57, 722, 1934, pp.218–23
- Davidson, C., 'The Yapp reflector of the Royal Observatory, Greenwich', *The Observatory*, 57, 720, 1934, pp.159–63
- Dowell, J. H., 'Slitless spectrograph for the Greenwich Observatory 36-in. reflector', *Journal of Scientific Instruments*, 12, 7, 1935, pp.224–5
- Greaves, W.M.H., Davidson, C. and E. Martin, 'The Zero Point of the Greenwich Colour Temperature System', *Monthly Notices of the Royal Astronomical Society*, 94, 6, 1932, pp.488–507
- Spencer Jones, Harold, *Annual Report of the Board of Visitors*, 2 June 1934, p.4

### 97. Occultation machine
- Pratt, Alex R., 'The Occultation Machine of HM Nautical Almanac Office', *Journal of the British Astronomical Association*, 124, 1, 2014, pp.12–21
- RGO Archives, 'The Prediction and Reduction of Occultations, 1937', RGO 16/355/2
- Wilkins, George A., 'The History of the H.M. Nautical Office', *Proceedings of the Nautical Almanac Office sesquicentennial symposium, U.S. Naval Observatory, March 3–4, 1999*, Alan D. Fiala and Steven J. Dick (eds), Washington, D.C. US Naval Observatory, pp.55–81

### 98. Philip Laurie's wartime diaries
- Blyzinsky, M., 'We Never Closed', *The Maritime Yearbook: Annual Magazine of the Friends of the National Maritime Museum*, no. 4, 1996/7, pp.40-2
- RGO Archives, 'Royal Observatory, Greenwich and R.G.O. History Papers (including Philip S. Laurie Papers), 1761–c.1982', RGO 107
✢ RGO Archives, 'Royal Observatory, Greenwich and R.G.O. History Papers (including Philip S. Laurie Papers), 1761–c.1982', RGO 107

### 99. Quartz clock and frequency standard
- Dolan, Graham, 'An introduction to the Quartz Clocks of the Greenwich Time Service', http://www.royalobservatorygreenwich.org/articles.php?article=1338 (accessed 29 August 2023).
- Smith, F, 'Quartz clocks of the Greenwich Time Service', *Monthly Notices of the Royal Astronomical Society*, 113, 1953, pp.67–80

- Spencer Jones, Harold, *Annual Report to the Board of Visitors*, 5 June 1937, p.37

**100. HP5061A caesium-beam atomic clock**
- Jones, T., *Splitting the Second: The Story of Atomic Time*, Institute of Physics Publishing, Bristol, 2000
- Smith, H., 'The steady march of atomic time', *New Scientist*, 93, 1292, 1982, pp.382–4
- Woolley, Richard van der Riet, *Annual Report to the Board of Visitors*, 6 June 1959, p.9
- Woolley, Richard van der Riet, 'Herstmonceux Atomic Standard and Greenwich Atomic Time Scale, G.A.', *Royal Observatory Bulletins*, 140, July–Sept 1966, p.B133

**Epilogue**
- Agence France-Presse in Paris, 'Do not adjust your clock: scientists call time on the leap second', *The Guardian*, 18 November 2022, https://www.theguardian.com/world/2022/nov/18/do-not-adjust-your-clock-scientists-call-time-on-the-leap-second (accessed 12 March 2025)
- Bennett, Jay, 'Is It Time to Redefine Time?', *Scientific American*, 18 February 2025, https://www.scientificamerican.com/article/worlds-most-accurate-clocks-could-redefine-time/ (accessed 17 March 2025)
- Bureau international des poids et mesures (BIPM), 'BIPM technical services: Time Metrology', https://www.bipm.org/en/time-metrology (accessed 12 March 2025)
- Dolan, Graham, 'The Herstmonceux years…1948–1990', http://www.royalobservatorygreenwich.org/articles.php?article=5 (accessed 13 January 2025)
- Dolan, Graham, 'WGS84 and the Greenwich Meridian', http://www.thegreenwichmeridian.org/tgm/articles.php?article=7 (accessed 12 March 2025)
- Littlewood, K. and Butler, B, *Of ships and stars: maritime heritage and the founding of the National Maritime Museum, Greenwich,* Athlone Press, London, 1998
- National Institute of Standards and Technology (NIST), 'How Do Atomic Clocks Work?', https://www.nist.gov/atomic-clocks/how-do-atomic-clocks-work (accessed 12 March 2025)

**Improvements in accuracy over time**
- Betts, Jonathan, *John Harrison and the Quest for Longitude* (2nd edition), National Maritime Museum, London, 2023, p.98
- Boucheron, Pierre H., 'Just How Good Was the Shortt Clock?', *NAWCC Bulletin*, 235, 22, 2, April 1985, pp.165–73
- Howse, Derek, *Greenwich Time and the Longitude*, Philip Wilson, London, 1997, p.170
- Hunter, Alan, 'Astronomy', *Science Progress*, 33, 131, January 1939, pp.509–17
- National Institute of Standards and Technology (NIST), 'A Brief History of Atomic time', https://www.nist.gov/atomic-clocks/brief-history-atomic-time (accessed 28 April 2025)
- National Institute of Standards and Technology (NIST), 'Keeping Time at NIST', https://www.nist.gov/blogs/taking-measure/keeping-time-nist (accessed 28 April 2025)
- Spencer Jones, Harold, 'The Determination of Precise Time', *Annual Report of the Board of Regents of The Smithsonian Institution*, 1949, Washington D.C., 1950

# Acknowledgements

Written over several years, this book has benefited from the support and advice of so many friends and colleagues both at Greenwich and beyond. In the first instance, I'd like to thank my Publishing colleagues Kathleen Bloomfield, Amelia Collins and Louise Jarrold who transformed an immense array of text files and images into the colourful publication that you see today.

The beautiful photographs in this book were made possible by colleagues in Conservation and Photography who worked hard to ensure that every object looked its best, for which I'd like to thank Karen Jensen, Nicolas Yates, Aisling Macken, Bethia Varik, Josh Akin, Sam Rowland and Charlotte Kite. Tina Warner kindly provided advice on image rights while Nat Elston provided essential collections support. I was also assisted by colleagues in the Caird Library team, led by Gareth Bellis, who kindly helped with the printed materials.

For historical advice and helpful chats over coffee, I'd like to thank my curatorial colleagues, past and present: Emily Akkermans, Megan Barford, Daisy Chamberlain, Alex Grover, Erika Jones and Henry Roberts, along with Stuart Bligh and Helen Mears. Special thanks to Daisy for her sterling work on acquiring and cataloguing an important collection of historic Observatory photographs, some of which feature in this book. Thanks also to Katherine Moulds who helped with the acquisition of these and other historic photographs.

A huge vote of thanks must go to our Curatorial Volunteer, Mike Dryland, who generously gave so much of his time and expertise to diligently read every section. Your feedback and encouragement kept me going and I'm truly grateful for your support. This book also benefited from the wealth of knowledge held by our Research Fellows, Daniel Belteki and Lee Macdonald, along with our Curators Emeriti, most notably Gloria Clifton and Graham Dolan. It would have been impossible without Graham's meticulous research and encyclopaedic website.

Beyond Royal Museums Greenwich, I'd like to acknowledge the specialist technical advice provided by contacts in the Scientific Instrument Society and the Scientific Instrument Commission who helped me turn mystery items tucked away in museum stores into key objects that unlocked so many fascinating new stories. Similarly, conversations and discussions with international colleagues at the four academic workshops held in 2021–22 as part of the Observatory Sites and Networks since 1780 project, funded by UKRI-AHRC, provided me with valuable new ideas and perspectives.

I would also like to offer my sincere thanks to Dr Emma Saunders, RGO archivist at Cambridge University Library, who gave much practical support in terms of accessing materials and images but who also provided intellectual support by drawing my attention to lesser-known items that helped contextualise and enrich these stories further.

In particular, I'd like to acknowledge the individuals who kindly provided advice on specific objects and sections: Stewart Ash, Jonathan Betts, Bill Burns, Stephen Burt, John Hearnshaw, Emma Hill, Caitlin Homes and David Rooney. A special thanks to Catherine Heymans for her insightful foreword.

Finally, this book is the culmination of a lifetime of love and support from my parents, Raymond and Margaret, along with many years of happiness with my husband Damian. You are the centre of my Universe: this book is sincerely dedicated to you all.

## IMAGE CREDITS

Object numbers for objects in the collection of the National Maritime Museum have been provided in captions and can be used to search for an object or artwork online at rmg.co.uk/collections

The publisher would like to thank the copyright holders for granting permission to reproduce the images illustrated. Every attempt has been made to trace accurate ownership of copyrighted images in this book. Any errors or omissions will be corrected in subsequent editions provided notification is sent to the publisher. Unless otherwise stated, images are © National Maritime Museum, Greenwich, London.

pp.12 (centre right), 106 © Thomas Adams / Courtesy of St Margaret's Church, Lee

p.13 (top middle) Australian National University Archives: Office of the Registrar, ANUA-16-176, detail of Sir Leonard Huxley, ANU Vice Chancellor, and Richard van der Riet Woolley.

p.13 (centre middle) Anne-Katrin Purkiss / The Royal Society

p.13 (centre right) © House of Lords / photography by Roger Harris. Attribution 3.0 (CC BY 3.0) licence, https://creativecommons.org/licenses/by/3.0/

p.13 (bottom) Amanda Clark / Cabinet Office

p.33 Longitude Act, 1714. Parliamentary Archives, HL/PO/PU/1/1713/13An35

p.45 Courtesy of Linda Hall Library of Science, Engineering & Technology. Attribution 4.0 International (CC BY 4.0) licence, https://creativecommons.org/licenses/by/4.0/

pp.70-71, 76-77, 95 © National Maritime Museum, Greenwich, London. Accepted by HM Government in lieu of inheritance Tax and allocated to the National Maritime Museum, 2012

pp.132, 141 Reproduced by kind permission of the Syndics of Cambridge University Library

pp.143, 192, 197 (right), 198, 212, 213, 218, 239 © Crown copyright. National Maritime Museum, Greenwich, London

p.145 British Geological Survey © UKRI. All rights reserved (BGS permit no. CP25/023)

p.152 Image provided by the National Meteorological Library and Archive, Exeter, UK. Royal Meteorological Society Collection

p.172 (right) CNUM - Conservatoire numérique des Arts et Métiers

p.195 Courtesy of the Vatican Observatory

pp.200, 201 (left), 219, 261, 262 Graham Dolan

p.201 (right), p. 240 (right) © The Board of Trustees of the Science Museum

pp.206-07 © National Maritime Museum, Greenwich, London. Courtesy of Steve Wates

pp.217, 255 Photographer: Scott Maloney / Cambridge University Library

p.243 TCB 417/E 80186, Courtesy of BT Group Archives

p.244 Reproduced with the permission of Sarah Allan

p.263 (left) © TfL from the London Transport Museum collection

---

First published in 2025 by Royal Museums Greenwich
Park Row, Greenwich, London, SE10 9NF
publishing@rmg.co.uk

ISBN: 978-1-906367-91-6

Text © National Maritime Museum, Greenwich, London
Louise Devoy has asserted her right under the Copyright, Designs and Patent Act 1988 to be identified as the author of this work.

At the heart of the UNESCO World Heritage Site of Maritime Greenwich are the four world-class attractions of Royal Museums Greenwich - the National Maritime Museum, the Royal Observatory, the Queen's House and Cutty Sark.

rmg.co.uk

All rights reserved. No part of this publication may be reproduced, stored in or introduced into a retrieval system, or transmitted in any form, or by any means (electronic, mechanical, photocopying, recording or otherwise) without the prior written permission of the publisher. Any person who commits any unauthorised act in relation to this publication may be liable to criminal prosecution and civil claims for damages. A CIP catalogue record for this book is available from the British Library.

Design by Ocky Murray
Printed and bound by Green Leaf Production, Slovenia

10 9 8 7 6 5 4 3 2 1

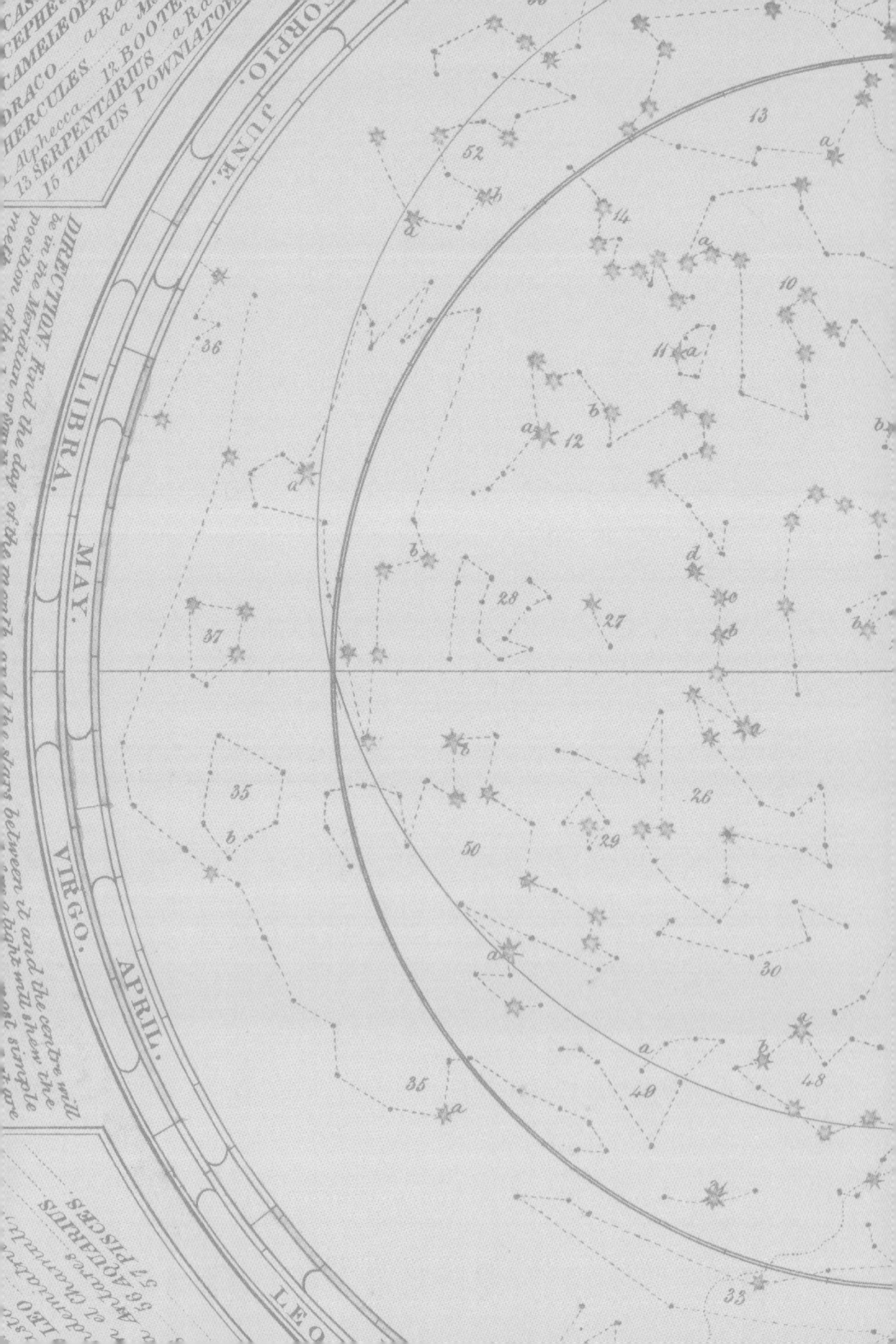